EMOTIONAL STRESS IN MONKEYS

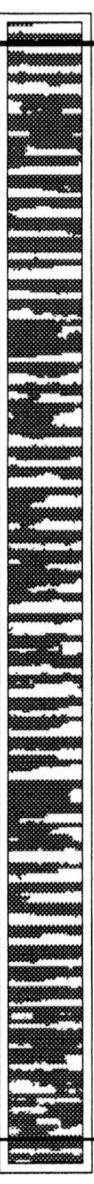

EMOTIONAL STRESS IN MONKEYS

by
A.M. Chirkov,
S.K. Chirkova
and M.G. Tsuliya

Translated by
Jeanne E. Farrow

NOVA SCIENCE PUBLISHERS, INC.

Art Director: Christopher Concannon
Graphics: Elenor Kallberg and Maria Ester Hawrys
Manuscript Coordinator: Roseann Pena
Book Production: Tammy Sauter, and Michelle Lalo
Circulation: Irene Kwartiroff and Annette Hellinger

Library of Congress Cataloging-in-Publication Data

Chirkov, A. M. (Aleksei Modestovich
    Emotional stress in monkeys / A.M. Chirkov,
S.K. Chirkova, M.G. Tsuliya
    p. cm.
    Includes bibliographical references and index.
    ISBN 1-56072-239-8 (hbk.) : $59.00
    1. Stress (Physiology)--Animals models.
2. Stress (Psychology)--Physiology. I. Chirkova, S. K.
(Svetlana Konstantinova) II. Tsuliya, M. G.
QP82.2.S8C538   1995   95-21127
619'.98--dc20   CIP

© 1995 Nova Science Publishers, Inc.
    6080 Jericho Turnpike, Suite 207
    Commack, New York 11725
    Tele. 516-499-3103 Fax 516-499-3146
    E Mail Novasci1@aol.com

All rights reserved. No part of this book may be reproduced, stored in a retrieval system or transmitted in any form or by any means: electronic, electrostatic, magnetic, tape, mechanical, photocopying, recording or otherwise without permission from the publishers.

Printed in the United States of America

CONTENTS

List of Abbreviations vi

Introduction 1

Chapter 1
The Modelling of Emotional Stress
and Neurogenic Illnesses in
Experiments On Monkeys 7

Chapter 2
The Activity of the Sympatho-Adrenal
System Under Emotional Stress
in Monkeys 21

Chapter 3
Gluco-Corticoid Hormones in Monkeys
Under Emotional Stress 51

Chapter 4
The Nature of the Change of the Endocrine
Function of the Testes in Monkeys
Under Conditions of Emotional Stress 79

Chapter 5
The Neuroendocrine Components
of Emotional Stress in Sexually
Immature Monkeys 95

Chapter 6
The Humoral-Hormonal Relationship
Under Emotional Stress in Monkeys 115

Chapter 7
Correction of the Neuroendocrine Impulses
Under Emotional Stress in Monkeys
with the Help of Pyrroxan 135

Conclusion 143

Bibliography 157

Subject Index 175

LIST OF ABBREVIATIONS

A	adrenaline
AG	arterial hypertension
AD	arterial pressure
VMK	vanilil-phenyl-glycolic acid
VND	higher nervous activity
GAMK	gamma-amino-butyric acid
GGAKS	hypothalamus-pituitary-adreno-cortical system
GGGS	hypothalamus-pituitary-gonad system
DA	3.4-dihydroxy phenyl ethylamine (dofamine)
KA	catecholamines
KRF	corticotropine-releasing factor (corticoliberin)
LG	luteinizing hormone
LG-RG	releasing factor for a luteinizing hormone (luliberin)
NA	norepinephrine
SAS	sympatho-adrenal system
FSG	follicle-stimulating hormone
CNS	central nervous system
CHSR	frequency of heart rhythm
CKG	chorionic gonadotropine of man
EKG	electrocardiogram
ES	emotional stress

INTRODUCTION

The problem of emotional stress is not by chance one of the most important problems in modern biology and medicine. Frequent and prolonged psycho-emotional overloads due to intense intellectual and physical activity, the expansive flow of information, working in shifts and under extreme conditions, man's flight into the cosmos and his pioneering new regions of the far north, a rapidly growing technology, the speed and intensification of production, urbanization, objective difficulties encountered while resolving new socio-economic problems, the breakdown of old traditions, the complication of interpersonal relations and many other factors create an objective predisposition for an increase in emotional tension in modern man, perceived by many authors as one of the leading causes of the emergence of a wide variety of psychological, nervous and psychosomatic illnesses. Therefore, the problem of emotional stress has not only a medical significance, but an economic significance, as well.

In recent years a certain degree of success has been achieved in the study of many medical-biological and socio-psychological aspects of emotional stress. We must note that a successful resolution of this problem is first and foremost based on the achievement of modern neuro-physiology and neuro-psychology, which have significantly expanded the theoretical understanding of emotion, of the mechanism of their display and the role of the emotional-stress reaction in the formation of various psycho-pathological and psychosomatic disturbances. The exceptional complexity and dynamism of the mechanisms of psycho-emotional reaction create significant difficulties in the study of human psychosomatic interrelations in clinic, which to a certain degree can be overcome with the help of a multi-discipline approach, including modern methods of psychoanalysis in conjunction with the study of the physiological and biochemical processes on various levels of the organization of the human functional system.

The research of N.P. Bekhtereva has been an exceptionally fruitful contribution to the development of the problem of neuro-physiological mechanisms of the creation of emotions. In this research a great number of fundamental movements of the physiology of the human brain were begun and developed, among them the principally new neuro-biological movement--the study of the brain code of psychic processes. For the first time in close cooperation with clinicians the research of the human emotional sphere was based on the knowledge of inner brain neurodynamics, which became possible under systematic observations of patients with live inner brain electrodes along with the wide use of a combined methodology approach to the study of the normality and mechanisms of human brain activity. A thorough generalization of numerous research works and an accumulation of facts dealing with the dynamics of long-term current pathological processes of the brain allowed N.P. Bekhtereva to propose and substantiate the theory of the emergence of a resistant pathological state, supported by the forming matrix of long-term memory, which created an opportunity to use new methods of treating brain diseases based on the activation of its hidden reserves, the formation of new branches of brain systems and the strengthening of brain defense mechanisms.

The analysis of the clinical and experimental date based on the theory of the functional system of P.K. Anokhin, the concept of the emotional excitement as a determinant of adaptive behavior, concepts of the fragmentary organization of emotional behavior and the informational theory of emotions, which fairly thoroughly expose the nature and the psycho-physiological mechanisms of the organization of emotional states, has important significance in understanding the role of emotions as biologically expedient adaptive factors of the integration of directed behavior acts in the broad picture of system reactions of the organism of man and animals. From the position of these concepts real opportunities open up, both for the thorough analysis of the state of the psycho-emotional sphere of man and the higher animals, the study of the sign of emotions, the degree of their expression, their quality and biological modality, and for the further development of the more fundamental movements in research on the problem of emotional stress with consideration for the neuro-dynamic, neuro-chemical, neuro-endocrine and somato-visceral correlators of the development of the stress-reactions, its stages, species and typological specificity of the psycho-emotional manifestations and the differences in the resistance of the organism to the action of psychogenic stimulants. Along with much material obtained as a result of the neurophysiological and neuro-chemical researches of the emotional-stress reactions, significant progress has been achieved in the field of psychopharmocology of emotional stress, the means to recognize it and the right criteria to evaluate the degree of expression. Cellular, molecular and sub-molecular mechanisms of stress are also being intensively studied.

One of the hopeful directions in resolving the problem of emotional stress and the clarification of pathogenetic mechanisms of psychosomatic illnesses is the study of the normalities in the functioning of the sympatho-adrenal and hypothalmus-hypopituitary-adrenocortical systems, the activation of which, as is well known, plays a leading role in the development of stress reactions and in the homeostatic mechanisms of regulating the organism's vital functions, as well as in the formation of the processes of adaptation on the basis of the mobilization of energetic and structural resources. In addition, numerous studies which have significantly expanded the concept of neuro-endocrine mechanisms of stress show that along with the emission of catechols and the increase in the level of gluco-corticoids in the blood, a normal display of stress-reaction is a change in the functional state of other regulatory neuro-chemical and neuro-endocrine systems, as well as the character of their relationship.

We may assume that one of the more hopeful criteria for the development of stress is the suppression of androgenopoez in sex glands. However, in spite of the biological adequacy of this reaction, lengthy and severe suppression of the endocrine function of the testes under the influence of intensive, lengthy or repeated stress stimulants can serve as the basis for various functional disturbances as a result of the "precipitation" of the physiological and metabolic effects of androgens and, first and foremost, lead to pathological changes in the reproductive system. At the same time the mechanisms of the suppression of the secretory activity of the gonads under stress have not yet been completely clarified.

Presently there is a significantly increased interest in the study of the biorhythmic and ontogenetic peculiarities of the hormonal manifestations of emotional stress, which allows us to hope for the creation of a single physiological platform for the generalization and accurate treatise of the numerous data obtained in neuro-endocrine research while studying the emotional-stress state. Research undertaken in this direction has allowed us to clarify the close interrelation of the phenomenon of the periodicity of the secretion of hormones and mediators with the nature of their response reactions to stress, on the one hand, and on the other--to show a certain dependence of these reactions on the stage of the organism's ontogenetic development. The study of daily and age peculiarities of change in the somato-visceral functions and the restructuring of the hormone balance under the severe and repeated action of psychogenic stimulants gives us the opportunity to thoroughly penetrate the phenomenon of stress and at the same time to approach a resolution of this complicated question of modern biology and medicine, the question of the specificity (selectivity) of the damage to the functional systems under emotional stress. Developing the concepts and principles of modeling selectivity of the violation of the visceral functions under emotional stress is determined to a great extent also by the adequate selection of the more "vulnerable" age period of the organism's development. In this, the complex studies of the hormonal activity of the leading branches of the neuro-endocrine system and visceral functions of the organism assume a particular relevance, which allows us to evaluate the character and variety of the forms of inter-endocrine and humorous-hormonal interrelations under stress, to reveal reserve possibilities and regulatory mechanisms of the neuro-endocrine compounds, a change in whose functional condition forestalls the development of structural changes in various organs and tissues, thereby assisting the achievement of resistance to the effect of unfavorable factors. This chiefly pertains to research on the neuro-endocrine mechanisms of the development of chronic and frequently repetitive emotional stress, inasmuch as many questions which touch on the hormonal security of these states remain so far unresolved. At the same time, it is under the chronic or frequent action of psycho-emotional stimulants that we see both more pronounced disturbances in the self-regulation of the function of the visceral systems, and an increase in the organism's adaptive ability.

The study of the basic regularities in restructuring the neuro-endocrine balance and the adaptive changes by internal secretion glands under emotional stress, as well as the clarifications of the nature of humorous-hormonal and inter-endocrine interrelations, is significant for developing a rational complex of prophylactic measures aimed at increasing the resistance of human and animal organisms to the action of intensive psychogenic stimulants. Revealing neuro-endocrine displacement under stress can also serve as an adequate basis for a search for new classes of biologically active agents--adaptagens, necessary to increase the organism's resistance under extreme conditions and under conditions of stress.

Experimental studies on the modelling of emotional-stress reactions in various laboratory animals make an enormous contribution to the development of the prospects of directions in the problem of emotional stress. We must note that for this purpose monkeys are the most adequate research object, due to their developed psycho-emotional sphere and the great morpho-functional similarity of the main physiological systems and the parameters of their neuro-endocrine regulation in man. Taking this into account, without attempting to solve the problem of the neuro-endocrine mechanisms of the development of emotional stress on the whole, we attempted to illuminate a number of questions devoted to the study of the function of the sympatho-adrenal, hypothalmus-hypopituitary-adrenocortical and gonad systems under the severe and repeated action of stress stimulants among the lower monkeys, to reveal the more hopeful paths to a solution of this problem, and to show concrete means for the evaluation and prophylactics of emotional stress.

The authors express their enormous gratitude to Professor N.P. Goncharov and candidate of medical sciences G.B. Katsia for the opportunity they offered to complete a large section of work on the determination of steroid hormones in the blood based on a laboratory of experimental endocrinology and consulting assistance in the discussion of the data obtained.

Chapter 1

THE MODELLING OF EMOTIONAL STRESS AND NEUROGENIC ILLNESSES IN EXPERIMENTS IN MONKEYS

One of the more important aspects of the problem of emotional stress (ES) is the development and creation of its adequate experimental models. The real basis for the reproduction and study of emotional states--varying in their biological modality, quality, degree of expression and signs--are modern theoretical concepts of the psychological and neurophysiological essence of emotions, their important role in the organization of the behavior of man and animal. In studies of recent years, along with a clarification of the reflecting-evaluating function of emotion, of the close tie with motivations, memory and consciousness, a similarity has been shown between the structural organization of the emotional zone in the brain of man and animal, which supports the possibility for objective study of the various forms of emotions in experiment on animals (Levy, 1972; Anokhin, 1975; Bekhterev, Smirnov, 1975; Startsev, 1976, 1977; Gubachev, et al., 1976; Bekhtereva, et al., 1978; Valdman, et al., 1979; Bekhtereva, 1980; Gubachev, Stabrovsky, 1981; Simonov, 1981; Vedyaev, Vorobieva, 1983; Kitaev-Smyk, 1983; Sokolov, Belova, 1983; Valdman, Poshivalov, 1984).

Inasmuch as emotions are, on the one hand, the subjective experiences of the individual, and on the other hand--the function of the nervous system, the prospects for studies of the psycho-physiological bases of emotional-stress reactions is closely tied to the development of a systematic approach to their analysis with the broad use of methodical opportunity of ethnology, neuropsychology, neurophysiology, neurochemistry, psychopharmocology, endocrinology and other disciplines. Very fruitful, both in building general concepts of emotional stress, and in developing various physiological and pathological models of emotional-stress reaction, has been the integrated approach to the study of emotional stress from the position of Anokhin's (1975) biological theory of emotions and theory of the functional system, which stipulates the need for the evaluation of individual neurophysiological, somatic, vegetative and other manifestations of emotional reaction in an unbroken tie with the integral architecture of the behavior act.

Emotions being the subjective experiences which arise in evolution as a result of the natural perfection of the nervous apparatus, according to Anokhin's (1975) biological theory of emotions, they are seen as a means for quick evaluation of the biological significance of an action of external and internal factors, needs of animals and their satisfaction. A significant addition to the essentially adaptive role of emotions in the process of adapting to external media, is the analysis of their physiological mechanism of including in the integral picture the behavior of man and animal from the position of the theory of the functional system of P.K. Anokhin. According to this theory, negative emotions arise when the results of a real action, aimed at satisfying a definite biological or social need, do not coincide with the previously programmed--expected--results, i.e., there occurs the so-called error of reverse afferentation on the results of the action with the property of the acceptor of the results of the action.

From the biological point of view negative emotions have an additional "energetic" charge, aimed at the active mobilization of the animal for repeated attempts to achieve the goal. The "coincidence" of reverse afferentation with the acceptor of the results of the action creates positive emotions, which sanction the success of the determined activity, i.e., they are the criteria for the sufficiency and usefulness of the completed behavior act (Sudakov, 1976, 1981). Simonov (1981) evaluates the physiological mechanism of emotions from a somewhat different position, considering that emotion is the active state of the system of specialized brain structures, prompting a behavior change in the direction of minimizing or maximizing these states. The informational theory of emotion developed by P.V. Simonov corresponds to this kind of physiological determination of emotions, perceiving emotions as the reflection (by the brain of higher animals and man) of any kind of active need and the appraisal of the likelihood of its satisfaction, based on innate or acquired experience. The low likelihood of the satisfaction of the subject's need results in negative emotions, which he actively minimizes, while an increase in the likelihood (possibilities) of the achievement of the result in comparison with the available prognosis leads to positive emotions, maximized by the subject.

In spite of well known difficulties in the treatment of individual patterns and aspects of animal behavior while evaluating the state of their psycho-emotional sphere, especially in extrapolating these data for man, the majority of researchers feel it possible to examine emotional reactions of animals as analogous to human emotion (Fedorov, 1977, Simonov, 1981; Vedyaev and Vorobieva, 1983).

According to Valdman and others (1979), a weighty basis for the expediency of modelling emotional stress and for reproducing basal processes (neurophysiological, neurochemical), which form the basis of his psychopathological manifestations in animals, is the need to recognize key, general biological mechanisms which participate in the regulation of psychic activity, which are fundamentally homologous in man and animal, since the organization of the elements of the functional system of the behavioral act--the neurophysiological basis or psychophysiological processes--is also singular. The significance of experimental study of emotional stresses in experiments on animals sharply increases if one considers that the reproduction of serious negative emotions in experiment on man is untenable, while their study in natural conditions is made difficult by ethical concepts (Fedorov, 1977).

To date a sufficiently large number of methodical means have been developed, which allow us to model emotional-stress states in animals with the reproduction of various biological types of behavior reactions (defense, aggression, food, sex, orienting-research), as well as certain psychopathological forms of behavior. It has been established that emotional stresses are formed in animals under conditions of immobilization, lengthy anticipation of harmful influences, constant irregular harmful or painful influences, electro-stimulation of negative emotional centers of the brain, the neuro-chemical activation of deep structures of the brain, hypokinesia, the activity of a large number of signals which have important information significance, and other stress stimulants (Cannon, 1927; Sudakov, 1976, 1979, 1981; Valdman, et al., 1979; Khomulo, 1982; Vedyaev and Vorobieva, 1983). We must note that under immobilization, of the activity of nosiseptive stimulants and other so-called physical factors, the trigger mechanism for the appearance of initial emotion-stress reaction is the perception and appraisal of the situation by animals as aversive, the result of the impossibility of avoiding it. Here it has been established that lengthy or frequently repeated conflict situations, producing a state of chronic emotional stress in animals in the experiment, leads to more significant neuro-physiological, somatic, visceral and hormonal displacements, in comparison with the effect of stress stimulants of short duration.

It is well known that under extreme emotional tension, exceeding the bounds of the adaptation of the brain system or, in other words, as a result of a breakdown in the adaptation mechanism, there appear various stress injuries to the organism, expressed in psychic and psychosomatic pathology. However, this aspect of emotional stress contains a great number of unresolved questions. In particular, it has not been totally explained why in some cases emotional stress leads to pronounced disturbances in the functions of various systems and organisms, with the development of persistent pathological states, whereas in other cases, it leads to the activation of the organism's natural anti-stress systems, achieving complete resistance. Especially interesting are the increased adaptive abilities of animals under short-term repeated emotional stress in those cases, when they are reproduced within equal periods of time or when they alternate with positive emotional influences (Sudakov, 1976, 1981; Meerson, 1981, 1984). Research of the mechanisms of resistance to emotional stress has become one of the actual problems of physiology (Sudakov, 1981).

It has been established that resistance to emotional stress is controlled by genetic and individual factors of development, and is determined by the peculiarities of the response reactions on the part of the central neuro-chemical systems of the brain, among which the important role in the appearance of negative emotions belongs to biogenic amines and neuro-peptides (Yumatov, 1983). The problem of resistance to emotional stress envisages the resolution of such questions as the study of species and age characteristics of stress, and it is closely tied to an explanation of the role of many attendant factors: the clarification of the relationship of the force of the stress stimulant and the state of the organism at the moment of interaction, the character and frequency of the action of stress stimulants, their coincidence to the phases of daily activity of the organism, etc..

The selection of the research object is very significant for the correct treatment of experimental data, including those data obtained in the study of physiological and pathological effects of stress, as well as the possibility for the extrapolation of these data to man. In the present stage of the development of primatology, there is no doubt that monkeys are of exceptional interest in studies exploring the physiological mechanisms of the development of stress, and consequent psychosomatic disturbances. The reason for this conclusion--and for the fact that the use of monkeys in medical-biological experiments in recent years has reached such enormous proportion--is seen in the data of many studies -- anthropological morpho-physiological, genetic, endocrine, biochemical and others -- which indicate an evolutionary proximity between man and monkey greater than had previously been assumed, and also a significant similarity in the fundamental physiological systems of man and monkey (Lapin, 1977; Friedman, 1977).

The high phylogenetic level of the structural-functional organization of a monkey's brain, reflected in the progressive development of certain brain formations in comparison with other animals, is manifested first and foremost in the qualitative and quantitative development of the new cortex, and especially the frontal lobes, the complication and organization of new receptor-cortical, cortical-subcortical and inter-cortical connections, which create a qualitatively new, more perfected integrative activity of the brain; it is also reflected in the more subtle differentiation and the significant complication of cyto-architectonic structures of the hypothalamus, hippocampus, cerebellum and other brain formations and inner brain connections, which, together with the greater similarity of the functional characteristics of cortical formation and sub-cortical-stem structures of the brain of man and monkey according to a great number of indicators, including the parameters of electrocorticograms and subcorticograms, attests to the irreplaceability of monkey as an experimental object for the further study and recognition of the physiological bases of psychic activity, of the normal functions of the central nervous system, of emotions and emotional stress, of neuroses and various somato-visceral disturbances (Urmancheyeva, 1972; Urmancheyeva, et al., 1977a, 1977b, 1977c, 1977d; Lapin, 1977; Cherkovich, Lapin, 1978; Khasabov, 1978; Cherkovich, et al., 1983).

The prospects for using monkeys for the study and resolution of various aspects of emotional stress is thus closely tied with the peculiarity of the structural-functional development of the primate brain, which explains the richness of the repertoire and of adaptive behavior, the variety of motion activity in monkeys, their exceptional manipulating ability, their high psycho-emotional reaction, the complex and dynamic group hierarchy relations and other personal forms of behavior, their significantly great ability--in comparison with other animals--for imitative activity and the great mobility of their nerve processes (Tikh, 1970; Firsov, 1982; Deryagina, et al., 1984; Lapin, et al., 1984; Butovskaya, et al., 1985, 1986). Together with the complication of the structure and function of the brain, the perfection of the support-movement apparatus, especially of the upper extremities, many authors note the great morpho-functional similarity between man and monkey in other physiological systems as well, including the cardiovascular and digestive systems (Startsev, 1971, 1972; Cherkovich, Fufacheva, 1973; Urmancheyeva, et al., 1977c; Lapin, 1977; Belkania, 1982; Cherkovich, et al., 1983; Belkania, Dartsmelia, 1984).

An exceptional similarity has been discovered between man and the higher--and certain lower--monkeys, especially Papio Hamadryas, in the function of the endocrine system, which is manifested in the similarity of many aspects of biosynthesis and the spectrum of produced hormones, neuro-endocrine mechanisms of internal gland secretion regulation, the antigen structure of tropic hormones of hypophysis, etc. (Goncharov, 1971; Chambers, Brown, 1976; Yudayev, et al., 1976; Goncharov, et al., 1977a, 1977b; Rose, et al., 1978; Plant, 1979; Tavadyan, 1981; Taranov, 1981; Katsia, et al., 1984a, 1984b; Chirkov, 1984; Tsulaya, 1985). We should note that emotional-stress reactions in monkeys, accompanied--as in other animals--by corresponding changes in the psycho-emotional sphere with their hormonal and somato-visceral provisions, are modified by characteristic--for primates, with their semi-vertical statics--peculiarities in the regulating many vital functions of the organism. This explains the similar direction of the reaction of the physiological systems and, in particular, the conformity of typological characteristics of the central and peripheral hemodynamics in man and monkey as the result of a single foundation for the mechanism of body orientation in space and the organization of a system reaction to the adaptation to existence in the Earth's gravitational field (Belkania, 1982; Belkania, Dartsmelia, 1984).

Without going into a detailed description of the means and methods used to neuroticize monkeys, we must emphasize that at the present time there are convincing data which support the effectiveness of modelling - on lower monkeys - neurogenic pathology, the clinical manifestation of which to a great extent corresponds to the nature of neurotic disturbances seen in man (Startsev, 1971, 1972, 1976, 1977; Lapin, Cherkovich, 1976; Lapin, 1977; Magakyan, 1977; Urmancheyeva, et al., 1977a, 1977b, 1977c, 1977d; Dzhalagonia, 1979). It has been established that emotional stresses, expressed neurotic and psychosomatic disturbances in monkeys, as in man, develop only in complex living and conflict situations, based on a disturbance of the normal family and group interrelations - on changes in the usual (customary) conditions and normal rhythm of life activity.

One of the more productive means for reproducing neurosis in lower monkeys--rhesus macaques and Papio Hamadryas--is the clash of natural conditioned and unconditioned reflexes, related to integral, system reactions of various biological forms of behavior: eating, sex, aggression-defense, family-hierarchic (communicative), orienteering-research, etc.. Other neuroticizing, emotional stress factors may be distortions of the daily rhythm of light, and feeding, an increased load on the analytical-synthesizing activity of the brain, especially in conjunction with defense reactions and negative emotions under group experiment conditions, etc. (Urmancheyeva, et al., 1977b, 1977c; Dzhalagonia, 1979). One fact draws attention to itself, that Papio Hamadryas command a high resistance of the higher neural activity processes to emotional-stress situations. This manifests itself in the quick "shifting" behavior away from unsolvable, conflict activity, to substitute activity. The ability to "shift" and the presence of expressed abilities in monkeys to adapt, creates a well known difficulty in creating neuroses, and the psychosomatic disturbances related to them, in these animals.

Dzhalagonia's research of many years (1979) on modelling and the study of experimental neuroses in monkeys has shown that factors inhibiting the realization of neurogenic influences are: the monotony of the nature of emotional-stress stimulants, the ability to quickly "shift" behavior to a substitute activity, loss of the ability to adequately perceive a situation as aversive after operative damage to the cortex of the frontal lobes, and each individual's strict regulation of its particular group-hierarchic position. At the same time, frequent changes in neuroticization in the absence of a clear pattern of stress influences, together with the influence of additional asthenic factors, lowers monkeys' adaptive ability and aids in the development of neurosis. These studies showed that one of the possible mechanisms for disturbing the higher nervous activity in monkeys under experimental neuroses can be the formation of pathological temporary connections between previous stereotypical activity of the animals, with a subsequent unavoidable activity of emotional stress (Dzhalagonia, 1979).

A detailed analysis of the background bioelectrical activity of the cortex and of various subcortical structures of the brain--the hippocamp, the amygdala and hypothalamus, which participate in regulating emotional behavior, as well as the study of electrical reactions of the structures studied to signals of various biological significance, conducted at various stages of the development of emotional stress and neurosis in monkeys, allowed us to establish that neurotic disturbances appearing as the result of chronic active psycho-emotional loads on higher nervous activity of primates is accompanied by constant discord between the neuronal systems of various levels of brain organization (Urmancheyeva, et al., 1977d). On the whole, changes in the higher nervous activity and vegeto-somatic functions in neurosis are expressed in the disturbance of the general and biologically specific forms of the behavior of monkeys, of various conditional reflex activity impulses, with irregular changes of the functional status of analytic systems, and also in the deregulation, and expression in various degrees, of the disturbances of individual cerebro-visceral functions and the metabolism (Lapin, Cherkovich, 1976; Magakyan, 1977; Urmancheyeva, et al., 1977a, 1977b, 1977c, 1977d; Dzhalagonia, 1979). Here the developing pathological changes of the function of visceral systems bears a persistent nature, and their normalization in the postneurotic period occurs, as a rule, later than the restoration of higher nervous activity.

In touching on the problem of the specific disturbances in somatovisceral function during emotional stress and neurosis, we must focus on certain questions concerning the influence strong emotions have on the activity of various physiological systems of the organism. As is well known, the development of emotional stress is determined by the inclusion of the central nervous and neuro-humorous regulating mechanisms - which integrate the organism's complete reaction - the development of which, in turn, leads to numerous non-specific impulses in a wide variety of systems and organs (cardio-vascular, digestive, hematogenic, support-motion, etc.), including changes in hydrogen, fat, protein and mineral exchange, the mobilization of the organism's structural and energetic resources. At the same time, being the cause of the disturbance of the physiological system's self-regulation, emotional stress (as shown in many studies) is far from always accompanied by the development of persistent and irreversible disturbances of the vegeto-somatic functions, and in certain circumstances can even appear as an adapting factor, stimulating psychic activity and contributing to a weakening and removal of previously formed persistent deregulations (Sudakov, 1976, 1981; Startsev, 1977; Kitaev-Smyk, 1983).

The high level of structural-functional organization, the reliability and flexibility of the central nervous mechanism regulating somato-visceral functions in monkeys, create significant difficulties for modelling neurogenic pathologies of the visceral systems. It has been established that under the influence of the widest variety of stress factors, including intensive psychogenic stimulants, in the majority of cases in monkeys only transitory changes in the function of the cardiovascular system are reproduced, with the development of reactive hypertensive states, reversible disturbances of stomach secretions and motility, etc. (Starstev, 1972; Belkania, 1982).

As shown by the research of many years in the modelling of neurogenic illnesses in man in experiments on monkeys for the recreation of persistent forms of a disturbance of the function of somato-visceral systems, not only are the nature, intensity, length and frequency of stress stimulants important, but so is the functional condition of the systems and organs at the moment of the action of the stress factor (Starstev, 1971, 1972, 1976). According to the principle of selectivity of injury of the functional systems under emotional stress, formulated on the basis of many years of research, disturbances in the function are developed first and foremost in that system, the natural arousal of which precedes stress and is repeatedly broken off by it (Startsev, 1977). Thus, for example, the immobilization of hungry male Papio hamadryas, leading to a natural defensive reaction, elicits functional changes immediately in several systems and organs, leading to a change in stomach secretions, heart activity, insulary apparatus, conditional feeding reflex movements, etc.. However, subsequent to the stress functions of the systems are restored.

But that same immobilization, if food had preceded it, eliciting natural activation of the functional system of feeding behavior, assumes a specificity and together with it--a pathogenicity, leading to chronic disturbances of the conditional feeding reflexes and the development of achylia, and -- under repeated associations -- leads to irreversible morphofunctional changes of the stomach lining, even to polyposis, adenomatosis and chronic ulcers. At the same time, while becoming pathogenic in relation to the digestive system, this stress remains non-pathogenic for other functional systems. Using combinations of activation (hyperfunction) of the physiological system with the overwhelming effect of emotional stress, in experiments on monkeys the following neurogenic psychosomatic illnesses have been intentionally successfully reproduced: arterial hypertension, stomach achylia and steroid-type paralyses and hyperkinesia, amenorrhea and impotence, and a diabetic state (Startsev, 1971, 1972, 1976; Repin, Startsev, 1975; Startsev, Chirkov, 1977). Illustration No. 1 (pages 26-27) presents data on the methods and conditions for the reproduction of emotional stress in monkeys, on the non-specific changes in various physiological systems and on the pathogenic influence of emotional stress on somatovisceral functions. The principle of selectivity of injury to the functional systems under emotional stress and neurosis found its further development and confirmation in the work of Dzhalagonia (1979) on modelling experimental neurosis in monkeys. As an emotional-stress stimulant, in place of immobilization, these studies used disturbance of the group-sexual interrelations--removal and transfer of a female from one male to another--coinciding with feeding time. The group-sexual stress in these conditions lead to the development of persistent neurosis with primary disturbances in the system of feeding behavior.

Consistent with the principle of selectivity of the disturbance of the visceral systems under emotional stress are the data of Meerson (1981, 1984), obtained while studying the structural-cellular bases for supporting the dominating system in adaptation to stress factors. The author notes that when an organism enters new environmental conditions, there occurs a decrease in cellular populations in the previously dominating structures, and what's more, the resistance of these structures to the influence of additional stress stimulants turns out to be significantly reduced, and the presentation of high loads leads to a breakdown in the mechanisms of adaptation.

To a certain extent this principle, reflecting on the whole the general biological regularity of the system reaction of a live organism in relationship to its initial functional state, is traced also through analysis of the changes in the functional state of the leading units of the neuroendocrine system. Thus, for example, with a high initial activity of the hypothalamus-hypopituitary-gonad system in monkeys in response to the influence of stress stimulants, we observe a sharp depression of this system, expressed in the change of the functional state of the central hypothalamus-hypopituitary mechanism of the regulation of the sex glands function, the suppression of androgen in the testes and the reduction of the concentration of testosterone in the blood. The introduction of stress stimulants in the setting of a borderline low initial activity of the hypothalamus-hypopituitary-gonad complex after a brief period of its additional suppression leads, on the contrary, to a significant increase in the production of testosterone and a sharp rise in the level of this hormone in the blood (Tavadyan, 1981; Tavadyan, Goncharov, 1981).

The in-depth study of the nature of the adreno-cortical reaction in monkeys in response to a wide variety of stress factors--a change in natural living conditions, an increase or decrease in the environmental temperature, hunger, immobilization, physical loads, submission to experiment, disturbances of the group-sexual relations and other stimulants, including neuroticization, x-ray irradiation, the influence of infectious agents, etc.--has allowed us to establish that the activation of GGAKS in the initial stage of response reaction development is primarily caused by psycho-emotional stress, the decrease and removal of which reveal specific aspects of hormonal manifestations of various physiological states of the organism, depending on the nature of the stress and its biological significance (Goncharov, 1971; Mason, 1971, 1974; Goncharov, et al., 1977a). It is no accident that in these studies unique factual data were obtained, which is one of the brighter pages in the development of the study of the specificity of many stress reaction manifestations, as well as in the reevaluation of Selye's concept of general adaptation syndrome (Selye, 1937, 1960), inasmuch as it is monkeys who have great emotion, and a great similarity to man in the nature of the hormonal mechanism producing the stress states of the organism.

In addition, the high psychophysiological reactivity of lower monkeys, who stormily react to the influence of many situational, and anthropogenic, factors, creates the need to observe strict measures in creating standard conditions for conducting physiological experiments, especially during psychoendocrine research. In order to create adequate conditions for conducting experiments, and to remove the effect of a number of additional stress factors, related to conflict situations arising from a change in the group-hierarchy relationships and to the effect of various situational, anthropogenic and other stimulants, difficult to consider but invariably accompanying the confinement of monkeys in common living cells of the nursery, in our laboratory corpus of the Department of Physiology and Pathology of VND Scientific Research Institute for Experimental Pathology and Therapy (SRIEPT) of the Academy of Medical Sciences, we equipped special rooms with primatological metabolic cages.

Individual cages, equipped with a semi-automatic clamping device, allowed us to significantly reduce the time required for submitting animals to experiment and to remove the effect of the stress factors related to catching and transferring monkeys from the admission cages to special "clamping" cages, to fixate the animal for the purpose of administering IV's and other experimental procedures. The moveable structures we developed and the mechanism for hanging the primatalogical cages (Chirkov, 1984) allowed us to optimize technical conditions for conducting experiments, to quickly and freely produce a biomaterial enclosure, and significantly improve caring for the animals. Within the room standard temperature and light conditions were maintained. Throughout the entire experiment, the monkeys were on a standard feeding schedule and ration. Constantly throughout the entire period of the research the animals' overall somatic state was monitored, including the simultaneous recording of cardio-vascular system indicators.

Thus, confining the monkeys in individual metabolic cages in a room separate from the nursery allowed us to a significant degree to standardize conditions for conducting experiments, to negate the effect of additional stress factors ("interference") while maintaining the opportunity for audio and visual contact between monkeys within a group.

Before the start of the experiment the monkeys adapted to the setting and conditions for conducting the experiments during a 3 to 4 week period, as shown in the research of Tavadyan (1981) and Taranov (1981), sufficient to establish a normal hormone balance and daily rhythm of the secretory activity of the steroid producing glands, characteristic for lower monkeys. To suppress the aggressive-defensive excitement over the experiment setting blood drawing procedure, during the period the monkeys were subjected to false IV treatment 3 times a week. In order to negate the influence of seasonal fluctuations in the size of the content of hormones and mediators in the biological environment the main series of experiments on monkeys were conducted during the same season.

The development of the stress state was reached with the help of a strict 2-hour immobilization, which is for this species of monkey a powerful psycho-emotional stimulant (Startsev, 1971, 1972, 1976). In all experiments immobilization of the monkeys was carried out according to the standards on special shields, in a supine horizontal position with head fixed and limbs bound. The method of 2-hour immobilization, used as a stress stimulator, differs in its simplicity, demands no operative interference, is easily reproduced, allows us to realize a series of repeat stress actions and to carry out experiments simultaneously on a group of animals. Research was carried out on lower monkeys--male *Papio Hamadryas*, born and raised in the Sukhumi Primatalogical Center of the Academy of Medical Sciences SRIEPT. In all, the experiments used 54 clinically healthy, intact, sexually mature male *Papio Hamadryas*, 20-35 kg in mass, 8-12 years in age, and 5 prepubescent males of the same species 1.5-2 years of age, and 6-8kg in mass.

In order to study the dynamics of the change of the content of the biogenic amines and steroid hormones under conditions of emotional stress and its consequences, various experiment settings were used, which took into account the possibility of clarifying the relationship of response hormonal reactions to time of day when the stress stimulant was introduced, its length, frequency, and the length of intervals between individual actions, as well as the age of the animals. In order to clarify the nature of functional interrelations between the activity of SAS (sympatho-adrenal-system) and of steroid producing glands, the analysis of their reserve abilities and prophylactic means of neuroendocrine impulses under ES (emotional stress), a series of experiments was conducted with the introduction of the releasing factor to the luteinizing hormone of the hypopituitary (LG-RG), of the chorionic gonadotropin (chKHG), metopiron, L-DOPA and pirroxan.

The study of the function of steroid-producing glands based on the level of the content of gluco-corticoids and androgens in the blood was conducted with the help of the highly sensitive radio-immunological methods. The methods used in our work were adapted to the plasma of monkeys, were modified and developed under the guidance of Doctor of Medical Sciences, Professor N.P. Goncharov at the base of the laboratory of experimental endocrinology of the AMS SRIEPT (Goncharov, et al., 1977b, 1979; Katsia, et al., 1984a). Some advantages of radio-immunological methods, besides their ability to register minimum amounts of the substance being studied, their reproduceability and specificity, are also the reliability of this method and its high productivity. The activity of SAS in monkeys was studied according to the dynamics of the content of catecholamines (adrenalin, norepinephrine, and dofamine) in the daily urine amounts with the help of selective fluorimetric methods on the "hitachi" spectrofluorimeter MPF4. We adapted this method for monkey urine, and modified it by the application of absorption chromatography under pressure (Chirkov, 1984).

In spite of the fact that fluorimetric methods for determining catecholamines have well known drawbacks in comparison with radio-enzyme methods, the use of this method in our work met the tasks we had set, and the results obtained on the basis of its implementation give an objective and informative evaluation of the functional state of SAS, and highly correlate with other physiological and biochemical indicators. We must note, that the simultaneous study of the amount of the excretion of catecholamines (CA)--adrenalin (A), norepinephrine (NE), and dofamine (DA)--adequately reflect the sympatho-adrenal activity, and allow us to make a judgment about the change in the reactivity of the SAS, its reserve and synthetic abilities in man and animals under the influence of various stress factors (Kvetnansky, Mikulay, 1970; Kassil, et al., 1973; Korobova, et al., 1977; Vasiliev, 1981; Bolshakova, 1982; Vasiliev, Chugunov, 1983).

Furthermore, depending on the tasks of the research, the determination of catecholamines in the urine in a number of cases is more preferable, since the level of their content in the blood can be affected by a wide variety of factors (the blood draw procedure, a change in body position, etc.), which, along with the rapid elimination of catecholamines from the circulating blood as a result of their short half-life, significantly restricts the use of this method, especially in lengthy experiments (Limpkins, et al., 1980). It is well known that the level of NE excretion reflects the state of the mediator unit, of the size of the discharge of A-hormonal, and the simultaneous determination in the urine of vanilil-mandelic acid (VMA) increases the informativeness of the research and allows us to make certain conclusions regarding the intensity of the metabolism of catecholamines (Gubachev, Stabrovsky, 1981; Bolshakova, 1982).

Anticipating the presentation of our own data, and the data of literature in the field, about the peculiarities of neuroendocrine impulses under ES, which make up the subject of the following five chapters of this monograph, we feel it necessary to note that we conducted a well-rounded study of the nature of hormonal balance restructuring in monkeys. We took into account the simultaneous study, according to given indicators, of the functional activity of the sympatho-adrenal, hypothalamus-hypopituitary-adreno-cortical-gonad system on various levels of the experiment: before, during, and after the influence of stress stimulators.

Considering the relative paucity of available published data, in this field, on neuroendocrine mechanisms of emotional stress in monkeys, which are, as we conclude from what we have stated, irreplaceable objects in psycho-endocrine research, and also the contradiction of available data on both the nature of neuro-hormonal reactions under repeated stress stimulators and the functional interaction of individual hormones and mediator under stress, and considering the extreme relevance of this movement in experimental medicine, we set the following tasks:

1. To study the dynamics of the change of the functional activity of SAS, GGAKS, GGGS under severe emotional stress, in a daily sense, and an age sense.
2. To clarify the peculiarities of the functional state of the neuroendocrine systems in monkeys under lengthy and repeated, frequent stress stimulators with various lengths of intervals between individual applications.
3. To analyze the nature of the humorous-hormonal and inter-endocrine interrelationships under ES with the help of functional tests, and to evaluate the possibility of psycho-pharmacological correction of neuro-endocrine impulses under severe stress.
4. To reveal the more pronounced peculiarities of neuro-endocrine supply of repeated ES in sexually immature monkeys, and to

establish the possible role of hormonal imbalance in the mechanisms of the development of neurogenic arterial hypertension.

The resolution of these tasks has not only a general theoretical significance for the clarification of the developmental mechanisms of emotional stress and the processes of adaptation to repeated stress stimulators, but also in the study of the more general regularities of the activity of the leading units of the neuro-endocrine system and its role in the processes of forming experimental neurogenic pathology; it is also a significant foundation for further in-depth, intensified medical-biological research in the problem of emotional stress in experiments on monkeys.

Chapter 2

THE ACTIVITY OF THE SYMPATHO-ADRENAL SYSTEM UNDER EMOTIONAL STRESS IN MONKEYS

As a leading regulatory system, the sympatho-adrenal system has a primary significance in the organization of the organism's immediate adaptational reaction under stress stimulants, which is achieved thanks to the participation of catecholamines in the nervous and hormonal mobilization mechanisms of the organism's functions and energetic resources, the integration of metabolic processes, and the overall realization of the adaptive-trophic influences on all vital processes (Cannon, 1927; Orbeli, 1949; Shalyapina, 1979; Axelrod, Reisine, 1984). In spite of the many works devoted to the dynamics of sympatho-adrenal activity under stress, the interest of researchers in this problem is not declining. This is explained not only by the important role of catecholamines in the organism's adaptation to extreme stimulants, but also by the significance of the functional disturbances of the central and peripheral catecholamine systems in pathogenesis of nervous-psychic, cardio-vascular and other illnesses (Levy, 1972; Anokhina, 1975; Zavodskaya, et al., 1977; Berezkin, Tarasov, 1978; Gubachev, Stabrovsky, 1981; Zabrodin, 1982; Ganelina, Borisova, 1983; Markel, 1983; Sokolov, Belova, 1983; Axelrod, Reisine, 1984; Meerson, 1984; Chugunov, Vasilyev, 1984; Vasilyev, Chugunov, 1985).

Much literature is dedicated to the structure and function of the SAS, to the questions of regulating adrenergic mediation, including the processes of biosynthesis, inactivation, metabolism, KA reception, and to their change under stress (Parvez, Parvez, 1972, 1978; Katsnelson, Stabrovsky, 1975; Utyevsky, Osinskaya, 1977; Oommen, Balasubramanian, 1977; Damalz, et al., 1979; Ventura, et al., 1979; Divac, et al., 1980; Nozdrachev, Pushkarev, 1980; Azhipa, 1981; Kopin, 1982; Axelrod, Reisine, 1984). Furthermore, in recent years, thanks to modern analytical methods, great successes have been achieved in clarifying the localization of KA-ergic neurons in the CNS (central nervous system), the clarification of molecular synaptic transfer processes and the mechanisms of the activity of KA (Budantsev, 1976; Anisimov, 1979; Zile, Klusha, 1979; Legg, 1982; Panin, 1983). Likewise, the principles of evaluating SAS activity and the method of studying KA exchange in man and animals under various physiological states of the organism, as well as under stress, have received fairly detailed illumination (Levy, 1972; Kassil, et al., 1973; Kassil, 1975; Bolshakova, 1976; Lidberg, et al., 1976; Anisimov, 1979; Lennart, 1981; Manukhin, et al., 1981).

There is information about the content and regional distribution of KA, their predecessors, metabolites in the central nervous system and on the periphery, as well as enzymes taking part in the processes of biosynthesis and the metabolism of KA in various species of monkey (Hoffman, et al., 1976; Chiba, et al., 1977; Jacobowitz, MacLean, 1978; Perlov, et al., 1978, 1979; Brown, et al., 1979; Schofield, Everitt, 1981; Elchisak, et al., 1983). It has been established that according to a number of parameters (KA contents, activity of the enzyme systems, speed of biosynthesis, nature of the metabolism and reaction to the influence of stress stimulants), monkeys show a certain difference from small laboratory animals and at the same time a similarity to man (Murali, Radhakrishnan, 1979; Chiba, et al., 1977; Davis, et al., 1979; Kramer, McKinney, 1979; Perlov, et al., 1979). Studies on monkeys have shown the important role of the catecholaminergic systems in the regulation of complex forms of behavior and in the formation of emotional states: aggression, fear, depression (Davis, et al., 1979; Perlov, et al., 1979; Redmond, 1979, 1980; Redmond, Huang, 1979). However, the majority of the studies in this field have been conducted primarily on rhesus macaques, whereas we found no published works dedicated to the study of the state of SAS in Papio hamadryas under conditions of severe and chronic emotional stress.

At the present time available literature does contain information on the influence of KA on the physiological and metabolic processes in the organism, which confirms the theory of Cannon (1927) on the role of SAS in reactions of alarm, i.e., in the organism's functional mobilization for a life-and-death struggle, and the theory of L.A. Orbeli (1949) on the adaptive-trophic influence of the sympathetic nervous system. Catecholamines contribute to the excitement of the CNS, increase arterial pressure, strengthen and speed up contractions of the heart muscle, expand the coronary vessels, increase the work capacity of the heart and skeletal musculature, cause blood redistribution leading to the optimal supply of energetic substratus to the tissues; they expand the bronchi and strengthen pulmonary ventilation, stimulate the demand for oxygen and oxygenating processes in the tissues, ensure glycogen breakdown and glucose supply to the organism, increase the utilization of glucose in skeletal muscles, stimulate lipolysis and the excretion of free fatty-acids, raise body temperature, etc., thereby participating in the integration of the leading physiological processes directed at the homeostatic regulation of the functions (Miloslavsky, et al., 1971; Levy, 1972; Budantsev, 1976; Utyevsky, Osinskaya, 1977; Shalyapina, 1979; Panin, 1983).

Numerous studies have established the SAS's extreme sensitivity to the broadest variety of stress factors. Thus, the activation of the SAS in healthy persons is seen under the influence of various psychogenic stimulators, negative factors of the external environment, intense psycho-emotional activity, sports competitions, a change in geographic belt, etc. (Bolshakova, et al., 1972; Levy, 1972; Kassil, et al., 1973; Payu, 1975; Gubachev, et al., 1976; Korobova, et al., 1977; Menshikov, et al., 1977; Vasilyev, 1981; Sokolov, Belova, 1983; Uchakina, 1983; Tigranyan, 1985). Increased KA secretion in various experimental animals occurs under conditions of ES, elicited by immobilization as well as in pain and trauma, muscle stresses, hemorrhagia, cooling, etc. (Kvetnansky, et al., 1979; Mustafin, Sitdikov, 1980; Belova, Kvetnansky, 1981; Kvetnansky, et al., 1981; Manukhin, et al., 1981; Homulo, 1982; Lilly, et al., 1982; Tsulaya, et al., 1984b). Other psychogenic factors raising the activity of the SAS in monkeys are: disturbances in group interrelations, density, placing a more aggressive animal of the same species into the cage, the development of the flight reaction, immobilization and other neuroticizing factors, leading to the development of severe or chronic ES (Raab, Storz, 1976; Perlov, et al., 1979; Chirkova, 1982; Chirkov, 1984; Tsulaya, et al., 1984a; Bolshakova, et al., 1985; Chirkov, et al., 1986a).

Study of SAS activity in clinics has allowed us to determine a broad spectrum of this system's functional restructuring, including changes in KA secretion, an increase or decrease in the activity of KA-synthesizing and KA-metabolizing enzymes, a distortes reaction to functional tests and stress. These impulses are revealed under many psychic and nervous illnesses (Anokhina, 1975; Vasilyev, 1981; Gubachev, Stabrovsky, 1981; Airapetyants, Wein, 1982; Mezentseva, 1982; Vasilyev, Chugunov, 1985), Cardio-vascular Pathology (Miloslavsky, et al., 1971; Berezkina, Tarasov, 1978; Grigorieva, 1978; Ganelina, Borisova, 1983; Sokolov, Belova, 1983), Surgical Interferences (Halter, et al., 1977) and many other illnesses.

A great contribution to the revelation of the pathogenesis of psychosomatic illnesses is being made by experimental research on the study of SAS activity under ES, allowing deep penetration into the disturbance mechanisms of adrenergic mediation. It has been established that the central KA-ergic systems play an important role in preserving the self-regulation of the physiological functions under ES, and furthermore a change in the level of KA in certain structures of the brain under these conditions correlates to a deviation in the amount of arterial pressure, the development of stomach ulcers, etc. (Belova, Kvetnansky, 1981; Markel, 1983; Mezentseva, 1982). It has also been shown that species and individual traits determining a reduction in the organism's resistance to ES are related not only to differences in adrenergic mediation and a deficit in norepinephrine synthesis in the brain, but are caused by KA synthesis activity in marrow matter of the adrenals and of the sympathic nerve endings (Belova, Kvetnansky, 1981; Vasilyev, 1981; Kvetnansky, et al., 1981; Mezentseva, 1982; Vasilyev, Chugunov, 1985).

Under conditions of severe emotional-pain stress we note a generalized response of all SAS units and, furthermore, changes in KA content and in the intensity of the processes of synthesis and slowing up their neuron capture in various organs continued to increase even during the 24-hour period after termination of the stress action (Manukhin, et al., 1981). Exceptions are the hypothalamus, establishing a tie between the nervous and endocrine systems, and the adrenals, securing the hormone function of SAS. In these organs KA content is restored significantly earlier, reaching or even exceeding the control level by the end of the first 24-hour period after ES, which, in the opinion of the author, is explained by the need to establish immediately the activity of the hypothalamus and the marrow matter of the adrenals for the normal functioning of other sections of the SAS. Many researchers point to the increase of KA exchange in the nuclei of the hypothalamus and other structures of the brain under stress, establishing also an increase in the circulation speed of catecholamines, a change in the activity of KA-synthesizing and KA-metabolizing enzymes, of the sensitivity of the adrenal-receptors, etc. (Saavedra, et al., 1979; Stone, 1981; Belova, Kvetnansky, 1981; Anokhina, 1984). These changes and, in particular, the increase in the level of NE in the central adrenergic structures, have the leading role in the activation of peripheral sections of the SAS (Kassil, 1975).

In the development of the response reaction of SAS to stress stimulants there are 3 main phases of reaction:

1. Activation of the marrow layer of the adrenals and increase in adrenaline emission into the blood. The level of KA rises in other biological fluids and tissue. KA content in the adrenals does not change as a result of the complete compensation of its secretion by synthesis.
2. An increase in the KA level in the blood and tissues, against a background of the gradual reduction of the adrenaline content in the adrenals as a result of the predominance of secretion over synthesis.
3. The absence of an accretion, or a phase of reduction of the concentration of KA and a sharp fall in adrenaline in the adrenals (Kassil, Matlina, 1973; Kassil, 1980). These phases

reflect the more common normalities of the SAS reaction to extreme stimulants, not excluding the peculiarity of the manifestation of the response reaction of this system, caused by the species and age of the animals, by the initial SAS functional activity, by the nature of the stress stimulant, its intensity, duration, etc..

At the present time it has been established that the reaction of the SAS (as of other physiological systems) to various stress stimulants depends on the phase of the organism's daily activity (Ksents, Blinova, 1976; Blinova, et al., 1978; Vasilyev, 1981; Kazin, 1982; Ratge, et al., 1982; Ganelina, Borisova, 1983; Poppai, et al., 1984). Blinova, et al. (1978) discovered that when the influence of the stress stimulants coincide with the acrophase of the C-rhythm of KA secretion in dogs, a reduction takes place in the degree of SAS activation in comparison to its reaction to stress developing from a low initial level of KA content in the blood. The similar nature of the change in the sympatho-adrenal activity has been established in healthy persons, employed in jobs of intense nervous-emotional work, and in shift work (Kassil, et al., 1973; Vasilyev, Chugunov, 1985). While working at night, i.e., during the period of marked reduction in KA excretions, we note a high degree of SAS activation and more pronounced disturbances of the daily rhythm of KA excretion in comparison with the change in the sympatho-adrenal activity during analogous work in the daytime (Kassil, et al., 1973; Vasilyev, 1981).

A change in the daily rhythm of KA excretion was discovered also under intensive muscle loads and athletic competitions, under the influence of negative environmental factors, a change in geographic belt, etc. (Vasilyev, 1981; Uchakina, 1983). A lengthy disturbance of the daily rhythm of the SAS activity was noted in clinic among humans under a disturbance of the emotional sphere, neuroses, arterial hypertension, ulcer conditions, etc. (Ratge, et al., 1982; Ganelina, Borisova, 1983; Vasilyev, Chugunov, 1985). However, daily peculiarities in the change of hormonal reactivity under ES remain largely unexplained and, furthermore, data about the direction of the reaction of the visceral system toward stimulation of the adaptation processes of the organism or the pathology, depending on the timing of the influence of stress stimulants to the phases of C-rhythm of the activity of the vegetative function, are scarce and contradictory. At the same time in studies on monkeys it has been established that ES leads to more pronounced disturbances in the function of that physiological system which, at the moment of the activity of the stress stimulant, was activated with the help of a natural biological stimulant (Startsev, 1971, 1972, 1976, 1977). Therefore we conducted research on the functional state and reserve abilities of the SAS under conditions of severe ES during morning and evening hours, and also under a prolonged

Factors and conditions for the reproduction of emotional stress

Experimental neuroticizing factors:

- Immobilization.
- Chronic VND traumatization based on periodic nosiseptic applications and their expectation.
- Conflict situations in which there is a clash between the arousal of various biological functional systems: feeding, aggression-defense, sex, orientation-investigation.
- Disturbance of the daily rhythm of life activities.
- Increased loads on the processes of analysis and synthesis, in conjunction with conflict situations and flight reaction.
- Loads on the VND processes, against a background of the organism's aesthenizations - removal and irritation of various brain structures, x-ray, castration, sensitization with a foreign serum, injection with neurotropic preparations.
- Combination of natural biological irritants and the application of stress irritants - immobilization, disturbance of the sexual-hierarchical relations, etc.

Other factors:

- Change in the natural structural organization of the herd and group. Influence of situational factors, related to the life of monkeys in captivity.
- Change in the normal ecological and customary conditions of confinement: crowded conditions or isolation; fixation in primatological chairs.
- Disturbance of the sexual-hierarchical, and other forms of zoo-social behavior: removal of the young from the female, removal of the femal from the male and her placement with a rival; placement with a more agressive individual.
- The influence of anthropogenic factors.
- Submission of animals for research, and the conduction of experiments.
- Transportation and the influence of other psychogenic factors.
- Disturbance of the technology of confinement and work with the monkeys.

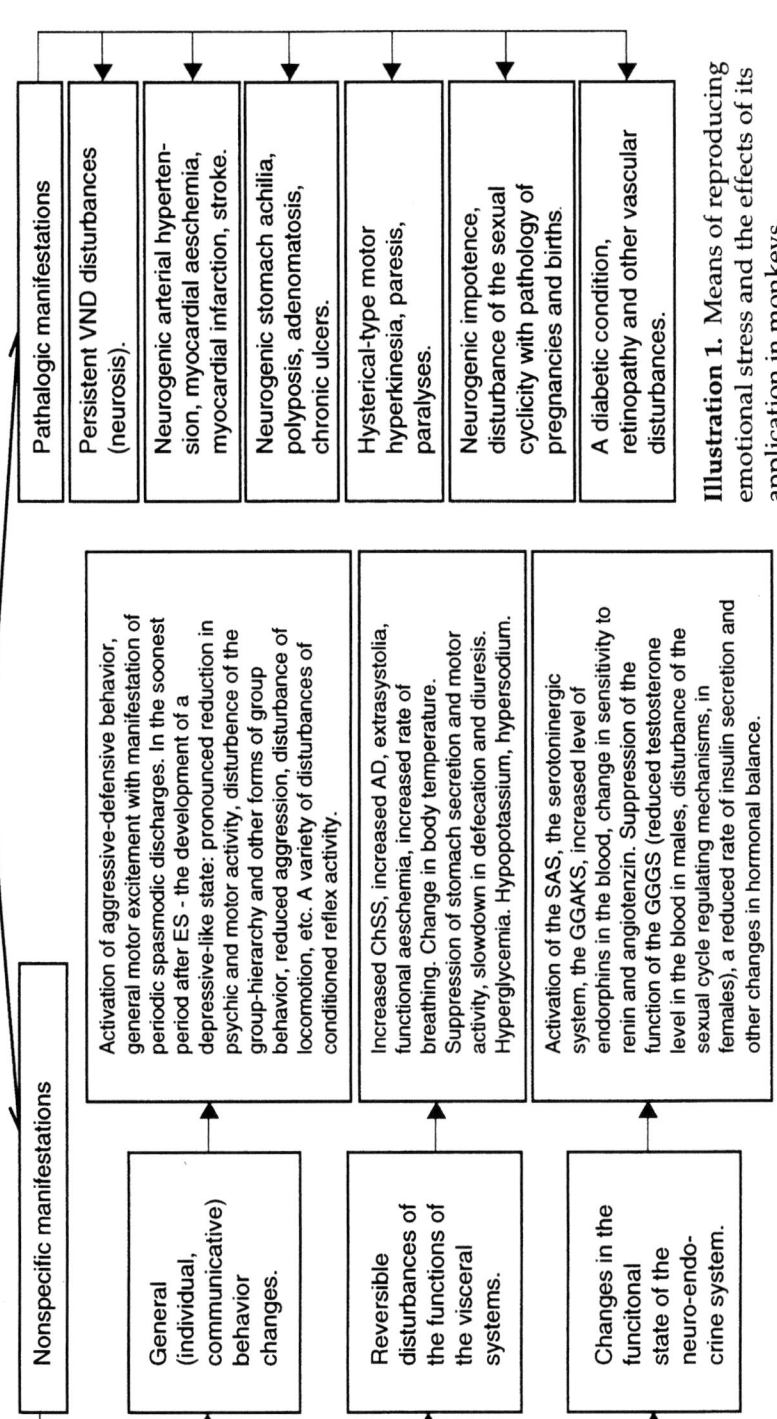

Illustration 1. Means of reproducing emotional stress and the effects of its application in monkeys.

stress stimulant (10-hour immobilization). In order to conduct the experiments on the nature of SAS function under the influence of stress stimulants it was necessary to first establish the standard (background) excretion amounts of free KA and VMK in male Papio hamadryas during their adaptation to confinement in individual metabolic cages. Throughout the monkeys' entire adaptation period to the experiment setting and conditions, over the course of 3 to 4 weeks, a daily collection was made of their daily amount of urine to determine the amount of A, NE, DA excretion. The VMK content was determined in daily quantities of urine collected during the first four 24-hour periods from the moment the monkeys were put into individual cages, and then, in the first 24-hour period of every following week.

Over the course of the one-week adaptation period, in accordance with how the monkeys adjusted to their new conditions of confinement, there occurred a gradual reduction in the amount of KA excretion (Illustration 1). Also by the third 24-hour period adrenaline excretions were 2.2 times less ($P<0.02$) than the amount of its discharge noted in the first 24-hour period of the monkeys' confinement in individual cages. The process of the reduction of NA excretion ended by the 6-8 24-hour periods, and its discharge during this period was half as much ($P<0.01$) in comparison with the first 24-hour period. The dynamics of the change of the DA excretion during the adaptation period were characterized by a less pronounced (statistically unreliable) and smoother reduction in the amount of the discharge. Stabilization of DA excretion was noted by the end of the second week. A reduction in the level of VMK excretion was completed by the fourth 24-hour period, and its amount was one third of the amount during the first 24-hour period of the monkeys' confinement.

Thus, during the period under study, there occurred among sexually mature monkeys a reduction in SAS activity, a high level of which was noted in the first 24-hour period of the monkeys' confinement in individual cages. The stabilization process of the amount of KA and VMK excretion allows us to conclude that the adaptation of SAS in research animals to the conditions of the experiment occurs over the course of the first 2 weeks.

Taking into account data from the literature on the VND normalization, the function of the visceral systems, the morphology of the blood and the biochemical indicators as related to the decrease in the aggressive-defensive arousal in monkeys (Utkin, 1960; Startsev, 1971, 1972, 1977; Kuksova, 1977; Dzhalagonia, 1979; Lemondzhava, Dzhalagonia, 1981), increased amounts of KA and VMK excretion in the first days of the monkeys' placement into individual cages can be seen as the development of a sufficiently pronounced emotional stress--a stress reaction in response to a change in setting and experimental procedures. The data obtained agree with the results of the studies

Table 1. Change in the daily excretion of adrenalin, norepinephrine and dofamin in the urine of sexually mature male Papio hamadryas (n = 20) under conditions of adjustment to confinement in individual cages.

Research Period (weeks)	Day of the week (24-hour)					
	first	second	third	fourth	fifth	sixth
Adrenalin (n-moles/24-hrs)						
1st	28.32±3.34	15.24±2.88 $P<0.05$	12.55±3.64 $P<0.02$	13.69±2.99 $P<0.02$	12.33±2.39 $P<0.01$	13.43±2.29 $P<0.01$
2nd	13.89±1.66 $P<0.01$	16.24±3.00 $P<0.05$	11.18±4.18 $P<0.02$	10.12±2.78 $P<0.01$	13.46±2.01 $P<0.01$	14.04±4.20 $P<0.05$
3rd	14.01±2.50 $P<0.01$	10.90±1.61 $P<0.01$	11.45±2.29 $P<0.01$	12.45±2.37 $P<0.01$	12.27±2.99 $P<0.01$	12.68±2.71 $P<0.01$
Norepinephrine (n-moles/24-hrs)						
1st	56.26±6.62	51.88±11.19	41.11±10.49 $P<0.05$	30.00±5.48 $P<0.02$	26.07±3.88 $P<0.01$	26.89±5.41 $P<0.01$
2nd	23.43±3.95 $P<0.01$	22.27±3.10 $P<0.002$	22.28±5.35 $P<0.01$	24.21±3.88 $P<0.01$	24.58±2.48 $P<0.01$	24.27±8.25 $P<0.02$
3rd	21.66±9.15 $P<0.02$	19.89±5.73 $P<0.01$	22.43±5.72 $P<0.01$	19.19±1.34 $P<0.001$	23.47±6.85 $P<0.01$	23.55±10.31 $P<0.05$
Dofamin (n-moles/24-hrs)						
1st	581.2±44.1	570.3±88.2	558.3±39.3	492.1±33.3	516.2±41.4	506.5±68.7
2nd	530.6±34.4	530.1±42.0	523.5±56.3	505.3±43.3	486.7±79.4	490.1±70.5
3rd	505.3±43.0	512.1±67.2	467.7±30.7	448.7±40.5	429.3±34.0	424.0±69.3

Note: P - reliability of the differences regarding the first days of the first week; absence of the mark P indicates the unreliability of the differences.

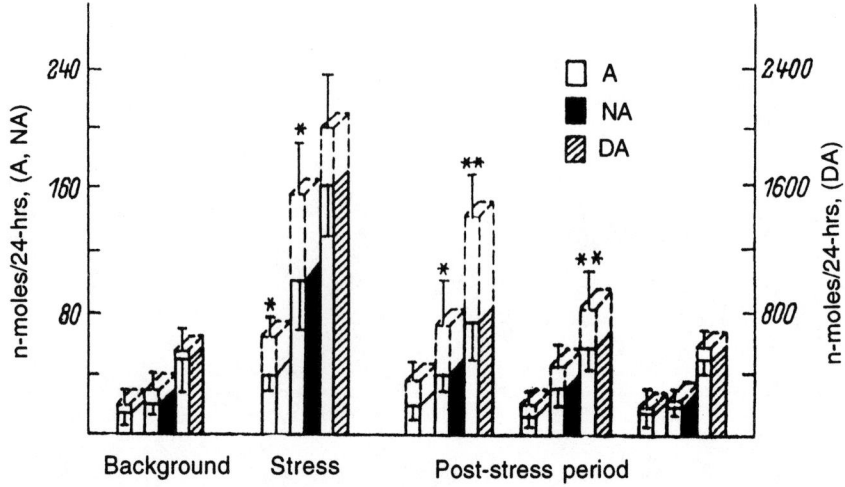

Illustration 2. Catecholamine excretion in urine among monkeys during emotional stress in morning and evening hours. Average arithmetical with reliable intervals.

Y-axis, on the right - excretion of dofamin (DA), n-moles/24-hrs; left - excretion of adrenalin (A) and norepinephrine (NA), n-moles/24-hrs; Solid line - stress in the morning (10:00 - 12:00); striped - evening stress (22:00-24:00). Reliability of the differences relative to the among of excretion under stress in the morning hours: * with $P<0.05$; ** - with $P<0.01$.

made on another species of monkey--rhesus macaque, in whom there was noted a gradual reduction in the level of excretion of the KA metabolites in urine, as they adjusted to permanent fixation in primatalogical chairs (Prolov, et al., 1979).

Illustration 2 presents data on the nature of the change of SAS activity in monkeys under extreme ES, reproduced in morning and evening hours. A 2-hour immobilization in the morning hours, i.e., in the phase of high motion activity, causes a sharp increase in KA excretion among sexually mature Papio hamadryas males. The level of A in urine increases during this time by 3.9 times ($P<0.01$), NA--by 5.1 times ($P<0.001$) and DA--by 3 times ($P<0.001$). Restoration of A and DA excretion was noted by the second day, and NA--by the third day of the period following stress. In percentage terms a greater increase in NA excretion under stress attested to the predominance of the activation of the SAS mediating unit over the hormonal unit--a coefficient of NA/A increased to 2.6 as opposed to 1.97 on the average. The increase in the coefficient NA/DA from 0.04 to 0.06 indicates the increase of KA biosynthesis. At the same time the increase in the size of the ratio (NA+A/DA) from 0.06 to 0.09 also allows us to assume a relative reduction in SAS reserve capabilities under conditions of severe ES, i.e., on a shortage of the initial product of the biosynthesis of KA-DOFA. However, in terms of percentages a significant increase in A, NE and DA by comparison with their initial amounts, indicates the sufficient expression of the synthetic abilities of SAS under these conditions.

The influence of an analogous stress stimulant in the evening (2200-2400 hours) led in the second group of monkeys to a greater SAS activation with a more lengthy period of the restoration of KA excretion to its initial values. Under these conditions excretion of A increased by 5.9 times ($P<0.001$), NE--by 7.1 times ($P<0.001$) and DA--by 3.5 times ($P<0.001$). The amounts of A, NE and DA excretion exceeded the level of their discharge under stress in morning hours correspondingly by 66% ($P<0.005$), 51% ($P<0.05$) and 20%.

A peculiarity of the SAS response reaction to stress during the evening was a more significant increase in A excretion, which was indicated by the lower value of the coefficient NA/A, amounting to 2.3 as opposed to 2.6 under stress during the morning hours. This allows us to conclude that during the evening immobilization stress proceeds with a greater activity of the hormonal (A) unit of SAS than in the morning hours. The coefficient of NA/DA under these conditions increased from 0.06 to 0.08, and the ratio (NA+A/DA) increased from 0.06 to 0.11.

The contrast of the change in the size of the ratio of summary KA under immobilization in morning and evening hours, and the greater increase - in terms of percentages - in DA excretions under stress during evening hours, indicate an adequate substrata and enzyme availability of the SAS function. The restoration of the excretion of KA and DOFA in monkeys subjected to immobilization during the evening occurred somewhat later, and by the second day the amounts of the discharge of A and NE were reliably higher than the initial values ($P<0.01$).

The sharp increase in SAS activity seen in male Papio hamadryas during a one-time two-hour immobilization attests to the development of severe emotional stress, which is also indicated by previously conducted studies, which established - in the given species of monkey under the influence of analogous stress stimulators - pronounced disturbances of higher nervous activity, of the cardio-vascular function and other visceral systems (Startsev, 1971; Startsev, Chirkov, 1977), as well as characteristic changes in the functional state of the cortex of the adrenals and the gonads (Goncharov, et al., 1978a). The increase in sympatho-adrenal activity established in our experiments agrees with available data on the increase in KA level in the blood and the increase of their excretion in urine in man and animals under ES

(Kvetnansky, Mikulay, 1970; Levy, 1972; Gubachev, et al., 1976; Malyshenko, 1976).

However, during immobilization of rats there is noted a primary activation of the SAS hormonal unit, which can be explained by a predominance of the content of A in the blood and in the adrenals in these animals under normal circumstances (Kvetnansky, Mikulay, 1970; Kvetnansky, et al., 1978; Davydova, 1978; Khomulo, 1982). The primary activation of the hormonal unit in rats under immobilization evidently lies at the basis of the formation of behavioral reactions with emotions of fear and upset, while in male Papio hamadryas under these conditions the elements of aggression and defense are more pronounced (Kvetnansky, Mikulay, 1970; Startsev, 1971).

The results of this experiment also witness the fact that the 2-hour immobilization conducted during the evening is a powerful neuro-emotional stimulant for male Papio hamadryas, eliciting a sharp increase in KA excretion. The observed changes in KA biosynthesis are realized under these conditions, just as under stress during the morning hours, against a background of sufficiently high reserves and synthetic abilities of SAS. The longer period needed for the restoration of the activity of the system to its initial functional state also corresponded to the large increase in the biosynthetic processes of SAS in experiments with a 2-hour immobilization during the evening. Our data agree with the research of other authors (Kassil, et al., 1973; Vassilyev, Chugunov, 1985), which shows that during a night shift people not adapted to night work show a more significant amount of KA excretion, a relative predominance of the activation of the hormonal (A) unit of SAS over the mediator (NE), an increase in the coefficient of the ratio of the sum of KA to DOFA (A+NE+DA/DOFA) and a "late" increase in KA excretion in comparison with analogous work during the day. The observed changes in KA biosynthesis under intense nervous-emotional work during the night is evaluated as a decrease in the reserve abilities of SAS, although our experiments conducted with monkeys produced data which was not convincing enough to support such a supposition.

According to modern theories, lengthy and frequently repeated stress situations lead to more pronounced changes in SAS function in comparison with the reaction of this system to the influence of short term stress stimulants. Furthermore, differences of the SAS response reactions are manifested according to a number of parameters: the amounts of secretion and the ratio among individual KA's, the change of their metabolism and inactivation, the speed of circulation, adreno-reception, etc. (Utevsky, Osinskaya, 1977; Bhagat, Subir, Chanda, 1979; U'Prichard, Kvetnansky, 1979). The more obvious dependence of the SAS reaction on the strength and duration of the stressor is presented in research on the activity of this system in man, under physical loads and during athletic competitions (Gubachev, et al., 1976; Korobova, et al., 1977; Vasilyev, 1981; Vasilyev, Chugunov, 1985).

In these studies it was determined that SAS activation is the result both of physical loads and emotional stress, more pronounced among athletes in the pre-start period and during competition. Severe training loads, especially athletic competitions, lead to activation of the adrenal and sympathic units of SAS, an increase of KA reserve (the increase of the discharge with urine of dofamin and DOFA), which is an adaptive reaction aimed at an increase in the organism of the active forms of hormones and mediators. The predominance of the activation of mediator (NA) unit over the hormonal (A) unit and a reduction of the increase in the level of KA excretion and their predecessors with urine, noted during frequent competition loads, in the opinion of the authors witness the economy of the function of SAS. These data agree with the studies of Menshikov, et al. (1977) and Bolshakova, et al. (1972), which established that the reduction of the secretory activity of SAS under chronic stress is one of the leading indicators of this system's adaptation.

Experiments on dogs have shown that under prolonged physical loads the concentration of A in the blood, which increases at the initial stage of exercise, was reduced during the period preceding the animals' complete exhaustion, while the content of NA increased (Brzezinska, et al., 1979).

The differences in the reaction of SAS to stress depending on the length of the stimulant are also observed on the level of adrenergic reception. Thus, under severe ES in rats there was established a slowdown in neuron capture and an increase in sensitivity of adreno-receptors to NE, which are seen as compensatory reactions, ensuring the increase of NE concentration in the synaptic fissure when there is a shortage of it, elicited by the stressor influence (Manukhin, et al., 1981). At the same time under chronic stress in rats there is noted a reduction of the sensitivity of adreno-receptors, which plays an important part in the processes of adaptation, and which prevents the development of depressive-type behavior reactions (Stone, 1981).

However, under extremely intense and lengthy stressor influences in man and animal one can see a pronounced decrease of the functional and reserve abilities of SAS, which is manifested in the absence of the KA increase to stress, in the exhaustion of NE tissue depots, in the suppression of biosynthesis and various disturbances of KA metabolism, which in the final analysis is the basis for the development of pathological processes (Anokhina, 1975; Sudakov, 1976; Zavodskaya, et al., 1977; Menshikov, Bolshakova, 1978; Vasilyev, 1981; Gubachev, Starbrovsky, 1981; Zabrodin, 1982; Khomulo, 1982; Vasilyev, Chugunov, 1983).

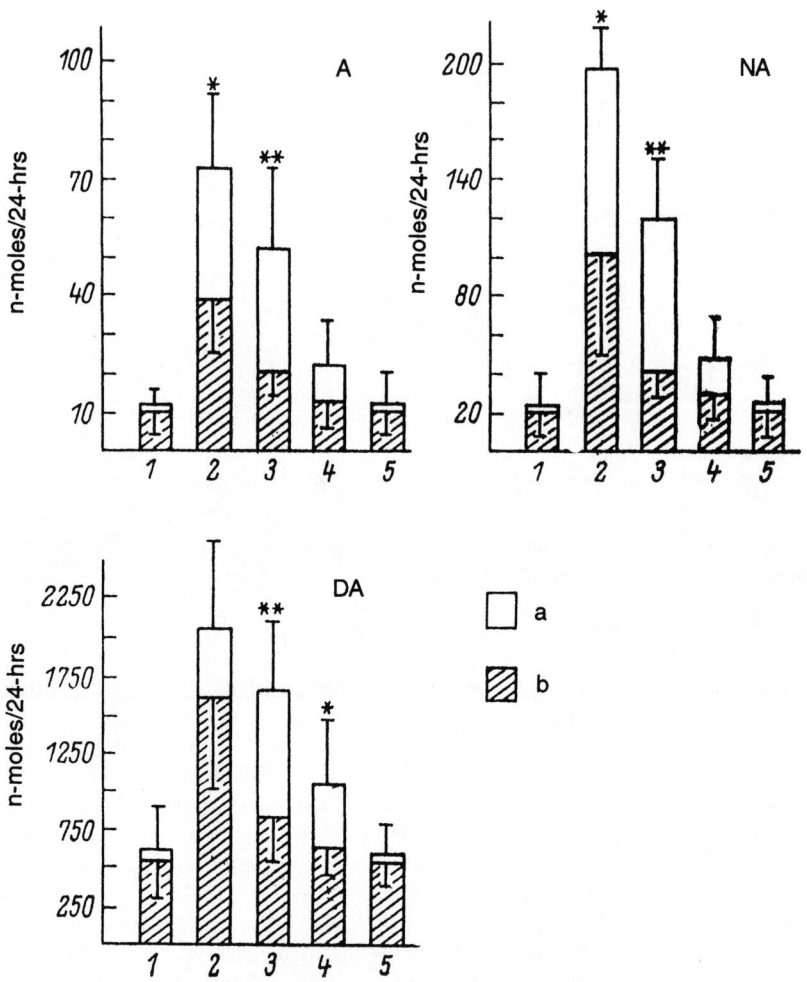

Illustration 3. Catecholamine excretion in urine of monkeys during a 10-hour immobilization

(a). Mathematical averages with reliable intervals.

Y-axis: amounts of catecholamine excretion, n-moles/24-hrs;
X-axis: 1 - background ; 2 - stress; 3,4,5, - respectively - the first, second and third 24-hour periods of the post-stress period; b - 2-hour immobilization. Reliability of the differences relative to the amounts of excretion during morning immobilization: * - with $P<0.05$; ** - with $P<0.01$.

As our studies have shown, a lengthy 10-hour immobilization (from 10:00 a.m. to 8:00 p.m.) in monkeys (the third experiment) elicits an even greater strain of SAS in comparison with the 2-hour immobilization conducted both during the morning and evening hours (Illustration 3). Under these conditions the discharge of A amounts to (72.96±12.9) n-moles/24 hours, which exceeds the initial value by 6.3 times ($P<0.002$), the excretion of NE increases by 8.6 times ($P<0.001$) and DA--by 3.9 times ($P<0.001$). Furthermore the amount of the excretion of A was 1.9 times ($P<0.05$), of NA--2 times ($P<0.05$) and DA--1.4 times higher by comparison with the corresponding amounts of their discharge in a 2-hour immobilization during morning hours. Less pronounced differences in the degree of SAS activation were received by comparison of the response reaction of this system to the influence of the stressor stimulant during evening hours. Thus, during the 10-hour immobilization the excretion of A, NE, and DA exceeded the corresponding amounts observed under a stress during the evening hours by only 1.2 ($P<0.05$), by 1.3 ($P<0.05$) and by 1.1 times.

We must note that during a 10-hour immobilization there was noted a predominance of the activity of the SAS mediator unit, witnessed by the increase of the coefficient NE/A up to 2.67 as opposed to the standard 1.87. For the duration of the immobilization there was also noted an increase in the coefficient NE + A/DA from 0.06 to 0.12. An analysis of these data, in a comparison with the nature of the response reaction of SAS to stress during evening hours, shows that during a lengthy emotional stress in monkeys there is noted a relative reduction in reserve abilities of the sympatho-adrenal system.

The restoration of KA excretion during the 10-hour immobilization occurred only by the 3rd day. The changes revealed in the activity of SAS during a 10-hour immobilization corresponded to the overall poor state of the animals after stress, which was expressed by a general slow-down, the absence of a reaction to humans or to food which was offered, constraint and unnatural poses, and a decrease in body temperature; furthermore, the expression of these manifestations significantly exceeded the behavior disturbances in the Gibbon species of monkey which were observed in experiments of a 2-hour immobilization (Butovskaya, et al., 1983; Deryagina, et al., 1984; Lapin, et al., 1984).

The results of our research attest to the development in monkeys, of a powerful neuro-emotional stress overload with escape into a depressive-type state, at the basis of which, evidently, lies the evacuation of the central and peripheral catacholamine depots, elicited by stress, which agrees with the data of available literature, indicating the important role of a shortage of endogenous KA in the development of depressive states in man (Anokhina, 1975; Mezentseva, 1982).

The increase in the degree of SAS activation during the increases length of immobilization up to 10 hours, established in our experiments, confirms numerous data in available literature about the SAS reaction's dependence on the degree and duration of extreme stimulants, with an increase in intensity of which there is observed a lag in the synthesis from secretions of KA in the adrenals, an exhaustion of the KA content in tissue, a reduction in reserve abilities of SAS, etc. (Bolshakova, et al., 1972; Zavodskaya, et al., 1977; Vasilyev, 1981; Zabrodin, 1982; Khomulo, 1982; Vasilyev, Chugunov, 1985). One of the more complex aspects concerning the neuro-hormonal mechanisms of the development of ES is the question of the nature of hormonal changes in man and animals under the influence of repeated stress stimulants. The resolution of this question anticipates the clarification of the general direction of the biosynthesis change and secretion of KA, the state of their inactivization and metabolic processes, the degree and physiological dependence of the predominance of the activation of SAS mediator or hormonal units, etc..

It is well known that the secretory activity and the predominance of the mediator or hormonal unit depend on many variables: the nature of the stimulant, its intensity, length, initial functional state, etc.. A large number of studies showing the dependence of the predominance of A or NE secretion on the nature of the stressor, species, age and individual differences, have been devoted to the matter of the specificity of the mediator and hormonal units' responses to various stressor stimuli (Sudakov, et al., 1979; Ismahan, et al., 1979; Popova, Koryakina, 1980; Belova, Kvetnansky, 1981; Kvetnansky, et al., 1981). Thus, in healthy males under an intellectual stress there were observed primary excretion of A, under a cooling effect--NE (Leblanc, 1979). At the same time, Bolotova, et al. (1980) discovered significant differences in the dynamic of KA excretion in healthy people to a single type of intellectual load depending on the degree of adaptation to active intellectual labor: In trained subjects in the pre-start period and 40 minutes after the load, NA discharge was predominant, in untrained subjects the primary excretion of A was revealed.

Various hypotheses exist to explain the nature of the interrelation between emotions and KA secretion (Levy, 1972; Lidberg, et al., 1976). Our authors' research confirms the well known thesis that the primary increase in the expulsion of A takes place in a state of alarm, whereas the expulsion of NE is related to active reactions of behavior (Kassil, 1975; Gubachev, et al., 1976). Furthermore, it has been shown that subjects with primary secretion of NE endure stress better. The majority of athletes participating in competition for the first time displayed in the pre-start period an increase in A excretion, while those who had adapted displayed NE. Kassil (1975) feels that stresses related to the development of such negative emotions as anticipation of pain, fear, horror, and alarm are accompanied by the predominant formation of adrenaline and its entrance into the inner medium, while those states demanding endurance, stamina, mental and physical load proceed against a background of increased NE secretion.

A certain specificity in the secretion of NE and A in response to psychogenic stimuli was revealed also in experiments on animals. Furthermore there was a close tie between aggressive behavior and primary secretion of NE, as well as a greater amount of its excretion among aggressive animals, in comparison with non-aggressive animals (Katsnelson, Stabrovsky, 1975; Miczek, 1981).

The change in the processes of adrenergic mediation in a blue dot has an important significance, as has been established in experiments on monkeys, in the development of emotional reactions of a negative character, in particular in the formation of emotions of fear and upset (Redmond, Huang, 1979; Redmond, 1980).

At the present time it is accepted that changes in KA exchange noted under stress, especially chronic stress, have an adaptive nature, reflecting the change in SAS function to a more economic regimen of reaction (Bolshakova, et al., 1972; Kassil, 1975; Menshikov, Bolshakova, 1978). In the studies of Payu (1975), in particular, it was shown that among athletes, while performing maximum-load tasks on the velo-ergometer, there occurs an increase of VMK excretion with urine, which happens against a background of the reduction of KA discharge with urine. Menshikov, et al. (1977) established that in healthy people during adaptation to conditions of the Far North, there occurs an increase in KA inactivization (an increase in the index VMK/A+NE), which simultaneously combines with the predominance of the secretory activity of the mediator unit (an increase of the coefficient NE/A).

In addition, contradictions have been discovered in analyzing the data of available literature on the change of the SAS reaction in response to the repeated influence of stressor-stimuli: Some authors point to the increase in the expulsion of KA, whereas others--to the decrease in the reactivity of this system. There is information which shows that the preliminary influence of stress increases the KA increase in blood plasma to the repeated stressor, while the speed of the increase of the KA level is in direct relation to the frequency of the applied stimulants (Kopin, et al., 1979). The preliminary immobilization of rats also significantly increased the increase of KA and the increase of the activity of dofamin-hydroxylase in response to subsequent decapitation (Kvetnansky, et al., 1978). Daily immobilization of rats for a period of 150 minutes for 40 and more days leads to an increase in the activity of KA-synthesizing enzymes (tyrosine-hydroxylase, dofamin-hydroxylase and phenylethanolamine-N-methyltransferase), an increase in KA production and their significantly increased secretion in urine (Kvetnansky, Mikulay, 1970; Kvetnansky, et al., 1970; Kvetnansky, et al., 1971; Kvetnansky, et al., 1978). In the studies of Khomulo (1982) emotional stress in rats (painful irritation by electric shock) was accompanied in repeated stressor actions by a gradual increase in KA discharge with urine: NE--from 0.17 to 0.69 MKG/day; A--from 0.08 to 0.29 MKG/day (by day number 130).

At the same time repeated electro-shock therapy in schizophrenics elicited a significant reduction in the increase of the NE level in blood plasma, in comparison with the increase of its concentration in response to the first application (Klimes, et al., 1979), whereas repeated loads (cooling and swimming) in mice lead to significant reduction of NE release from the sympathetic nerve endings (Benedict, et al., 1979).

The existing contradictions in SAS reaction can be explained by the various lengths of periods between stress influences and the frequency of their application. The latter is illustrated by the work of Rubanov and Riman (1980), who showed that the resistance of monoaminergic systems of the brain to "a hypoxic blow" can be increased by a 4-day application of frequent hypoxic loads, while an 8-day cycle of the same loads elicits a severe disturbance of KA exchange and other mediators, which is manifested in a significant increase of NE, serotonin and their predecessors as a result of the suppression of MAO activity. It has also been established that repeated ethereal stress with 30-minute intervals elicits significantly greater changes in the content of monoamines, including NE, in the various structures of the brain, in comparison with severe and prolonged stress (Teledgu, et al., 1979). Consequently, under the repeated application of stress stimulants there occurs a change in the response reaction of SAS on the periphery, conforming to the change in KA exchange in the various structures of the brain.

Inasmuch as the reaction of any physiological system, including the sympatho-adrenal system, to repeated stress stimuli is realized against the setting of its functional state, changing under the influence of the previous stimulant, it is logical to suppose that one of the important factors determining the formation of the general direction and degree of the expression of change of the reactivity of the regulatory systems under monotonous repetitive emotional stresses is the length of interval between application of individual psychogenic stimulants. In addition, the increased or decreased time interval between the application of the first, and repeated, stimulants, evidently leads to various quantitative and qualitative changes of the restorative processes of physiological functions with certain differences in the formation of "the structural trace," which is at the basis of the development of the adaptation of functional systems to the broadest variety of stress applications (Meerson, 1981, 1984). Thus, we can conclude that the nature of the response reaction of physiological systems of the organism to each concrete application of repeated stress stimulants depends not only on which phase of change of the functional state of the system it falls on in the restoration period, but is also determined simultaneously by the degree of development of adaptation processes, the inclusion of which was elicited by preliminary or preceding stress influences.

Considering the complex nature of the interaction of various neuromediator, humorous and hormonal regulatory systems in the organism's adaptation mechanisms, as well as the contradiction of available data on the change of the SAS function in man and animals under repeated stress and the possible role in these processes of the temporary application of stress stimulants and, in particular, the length of intervals between individual applications, it seemed expedient to study the dynamics of the neuro-hormonal indicators in monkeys under frequent stress situations, for the purpose of clarifying the possibility of the directed increase of the resistance of the neuro-endocrine system to repeated ES. In the concrete resolution of the question of the choice of experimental conditions and the time regimen for reproducing frequent ES in monkeys, we proceeded from available data about the possible development of adaptive processes on the part of VND in these animals to the action of various neuroticizing factors, especially in those cases when their application had a repeated and regular nature (Dzhalagonia, 1979). In addition, the repetitiveness of the regular stress applications was fundamental in modeling the neurogenic pathology of the somatovisceral systems in monkeys (Startsev, 1971, 1972, 1977), which also points to the need to clarify the concrete conditions and peculiarities of the structure of the time regimen of discrete emotional stress, leading in some instances to the development of disadaptation and pathology, and in other cases to an increase in the organism's resistance.

To study the nature of SAS activity under repeated ES, and to clarify the possibility of the increased resistance of SAS to its application, we conducted 3 series of experiments: The first series--a cycle of daily 2-hour immobilizations, produced for a period of 6 days (first group of monkeys); the second series--a similar cycle of stress applications with preliminary immobilization 3 days before the beginning of the cycle (second group of monkeys); the third series--a cycle of repeated 2-hour immobilizations with an increase in the intervals between individual applications, up to 72 hours (third group of monkeys).

The length of the interval between the preliminary stress application and the cycle of immobilization in the second series of experiments, and between individual immobilizations in the third series amounted to, correspondingly, 3- and 2-day periods, which, as the aforementioned data show, was sufficient to restore the functional

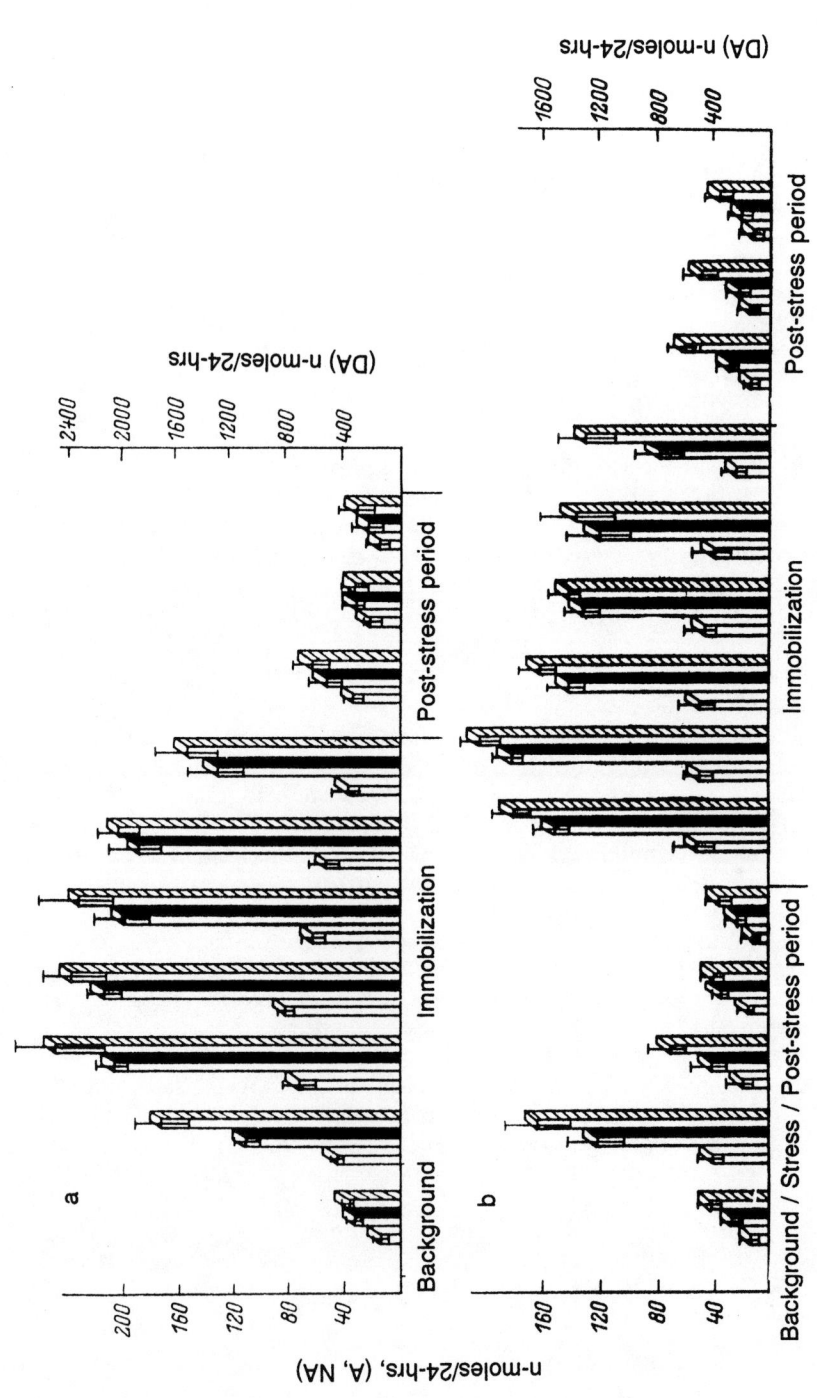

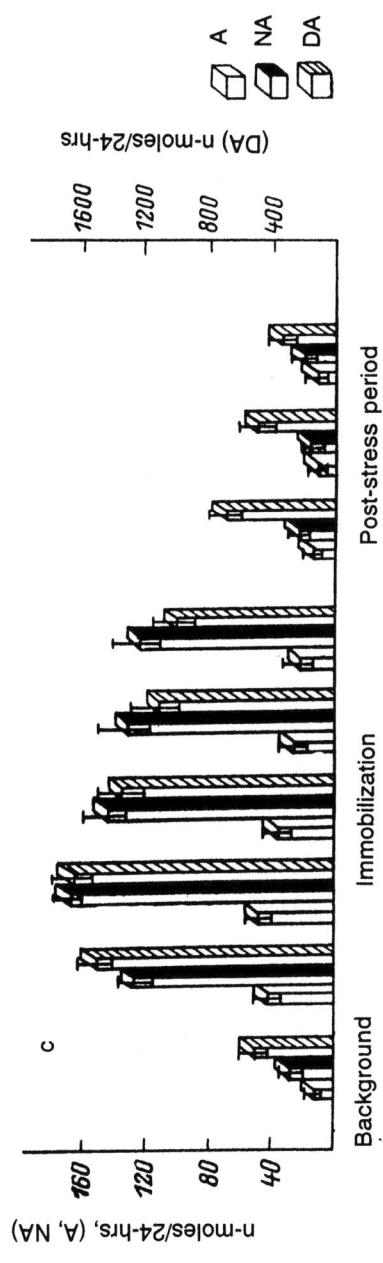

Illustration 4. Dynamics of catecholamine excretion in monkey urine with various regimens of the application of stress irritants. Mathematical averages with reliable intervals.

Y-axis: amounts of dofamine (DA) excretion, n-moles/24-hrs; left - excretion of adrenaline (A) and norepinephrine (NA), n-moles/24-hrs; a - cycle of daily immobilizations for a 6-day period; b - analogous cycle of stress applications with a preliminary (3-days prior) immobilization (stress); c - cycle of repeated immobilizations, separated by intervals of 2, 24-hour periods.

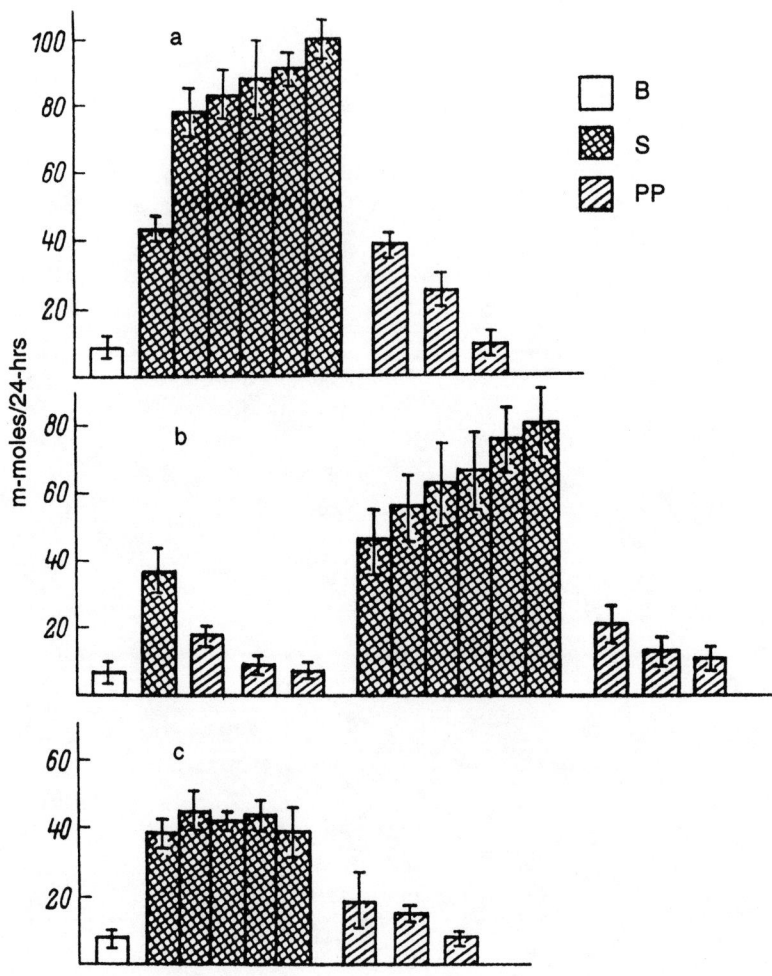

Illustration 5. Dynamics of vanilil-phenyl-glycolid acid in monkeys' urine various regimens of the application of stress irritants. Mathematical averages with reliable intervals.

Y-axis - Excretion of vanilil - phenyl-glycolic acid (VMK), m-moles/24-hrs; a - cycle of daily 2-hour immobilizations; b - analogous cycle of stress applications with preliminary (3 days prior) immobilization; c - cycle of repeated immobilizations, separated by intervals of 2, 24-hour periods. B - background; S - stress, PP - post-stress period.

SAS activity in monkeys after a single 2-hour immobilization. As has been established, daily 2-hour immobilizations conducted for a period of 6 days (first group), lead to a significantly greater activation of SAS in comparison with the pronounced reaction of this system to severe stress (Illustration 4, 5). The increase of KA and VMK excretion to the second immobilization significantly exceeded the increase of their level in response to the first application and amounted to, for A--(74.78±3.22) n-moles/24 hours ($P<0.0001$), for NE--(217.05±5.73) n-moles/24 hours ($P<0.0001$), and for DA--(1760.8±345.3) n-moles/24 hours ($P<0.0001$) and VMK--(77.17±2.88) n-moles/24 hours ($P<0.0001$). After reaching the maximum amounts of KA excretion, noted for A and NE on the third immobilization, and for DA on the second immobilization, we noted a gradual reduction in the stress level of their discharge (Illustration 4,a). At the same time the amounts of VMK excretion continued to increase in response to each subsequent application, reaching a maximum at the last immobilization (Illustration 5a).

The excretion of NA under repeated stress applications exceeded, in a percentage expression, the values of A excretion. Thus, in the period of their maximum discharge (the third immobilization) the increase of NA excretion amounted to 1,080%, whereas A--870%. The mediator-hormonal ratio under conditions of frequent application of stress stimulants increased more significantly in comparison with severe stress, reaching a value of 3.5 by the end of the stress cycle. The coefficient (VMK x 1000)/(A+NE) increased to 664. The normalization of KA and VMK discharge was reached by the third day.

Preliminary immobilization, conducted 3 days before the daily stress cycle, in monkeys of the second group led to a lower SAS activation (Illustration 4,b). The maximum increase of NE excretion in these animals was noted during the second immobilization and amounted to (19.51±10.64) n-moles/24 hours, but was 16% less than ($P<0.05$) the maximum amount of NE discharge in monkeys of the first group. The process of lowering the stress level of NE discharge in urine in these monkeys proceeded in a more pronounced manner, and by the final (sixth) immobilization the amount of its excretion was 1.5 times less than in animals of the first group under an analogous stress application.

The dynamics of DA discharge on the whole repeated the characteristics of the change of NE excretion, however, by the final stress application and the first two days of the follow-up period, the DA level in percentages exceeded the corresponding amounts of NE discharge. The nature of the change of the adrenaline level in response to repeated immobilizations in monkeys of the second group differed by the absence of a statistically reliable increase in the excretion of this hormone in comparison with its discharge by the first application.

During the entire stress cycle in these animals, just as in monkeys of the first group, there occurred an increase in the coefficient NE/A from 1.97 (standard) to 3.55 (sixth immobilization). In spite of the reduced amounts of VMK excretion of monkeys of the second group, the dynamics of the discharge of this metabolite on the whole repeated the increasing nature of its excretion among animals of the first group (Illustration 5b). The ratio (VMK x 1000)/(A+NE) was also high, especially in the second half of the stress cycle, and by the final immobilization the increase of this indicator as compared to its initial amount was 240%.

Additional immobilization conducted in monkeys of the second group by the seventh day of the post-stress cycle showed that an increase in the activity of the mediator and hormonal units of SAS and an increase in KA metabolism in these conditions were identical to the SAS reaction under severe stress in intact animals (Bolshakova, et al., 1985). Under repeated immobilizations, produced at intervals of 2 days, among the monkeys of the third group there was noted on the whole a similar direction in the dynamics of the excretion of A, NE and DA with the nature of their discharge in animals of the second group (Illustration 4,c). At the same time, as opposed to monkeys of the first two groups, among animals of the third group no increase was noted in the amounts of the excretion of the basic metabolite of catecholamines-VMK, the level of which in the daily urine, by the final immobilization ,was virtually no different from the reaction to the first stress application, and furthermore was 2 times lower ($P<0.01$) than in animals of the second group in the corresponding period of the study (Illustration 5,c).

In analyzing the data obtained about the activity of SAS in sexually mature Papio hamadryas males while adjusting to individual cages, under severe and repeated ES, we can come to the conclusion that the dynamics of the excretion of KA and VMK in monkeys, just as in other experimental animals and in man, is a very hopeful indicator, which adequately reflects the state of the psycho-emotional sphere from an insignificant degree of emotional tension to the development of pronounced emotional stress.

The sharp increase (10- to 12-fold) in KA injection and the 15-fold increase (by the sixth immobilization) of VMK excretion, noted during the daily application of stress stimulants, points to the extreme activation of SAS. A reduction in the stress level of KA excretion, beginning immediately after reaching maximum values, cannot be explained by the exhaustion of the catecholamines in the adrenergic nervous endings and in the medulla matter of the adrenals, since it occurred against the background of a sharp increase in VMK discharge. An increase in KA inactivation does not allow us to evaluate the reduction of the amounts of the response injection of A and NE as the result of a pronounced slowdown of the synthesis or a significant decrease of the secretory activity of SAS. The revealed changes of the excretion in urine of KA and VMK, in monkeys under repeated stress, bears a similarity to the dynamics of the response reaction under conditions of chronic stress in healthy people (Bolshakova, et al., 1972; Menshikov, Bolshakova, 1978).

These studies also revealed a decrease in the stress level of free KA excretion and an increase of their metabolism, accompanied by an increase in the level of VMK discharge, which, in the opinion of the authors, is a more characteristic sign of SAS adaptation to the influence of chronic stress. The dynamics of change of KA excretion, similar to our data, were noted among athletes during the performance of a series of similar repetitive loads in competitions (Payu, 1975; Gubachev, et al., 1976; Korobova, et al., 1977; Vasilyev, 1981). The gradual reduction of the degree of SAS activation, occurring parallel with the improvement of the athletes' performance, allowed the authors to make the conclusion about the economy of the functioning of the sympatho-adrenal system under conditions of repeated stress situations.

In connection with the foregoing, a decrease in the stress amounts of KA excretion to repeated immobilization in monkeys, in our opinion, should be seen as an adaptive change of this system's functioning under conditions of chronic stress. Furthermore the reduction of the degree of SAS activation can be compensated for by an increase in adrenosensitivity of the effector organs to NE (Manukhin, et al., 1981).

At the same time in the adrenergic structures of the brain under conditions of chronic stress there occurs a reduction in sensitivity to NE which, along with the change in the mediator exchange, is an adaptive process designed to increase resistance to stress and has a great similarity in its mechanism to the influence of anti-depressants (Stone, 1981; Platt, Stone, 1982; Saavedra, 1982).

However, experiments on rats in response to repeated stress applications revealed an increase in KA injection in the blood and an increase in A and NE excretion in urine (Kvetnansky, et al., 1970, 1971, 1978; Khomulo, 1982). This is possibly related to species differences of the SAS reaction to the influence of stress stimulants, as well as to the peculiarities of KA inactivation in small laboratory animals as opposed to KA metabolism in monkeys under stress.

In our opinion, what is important is the fact that preliminary immobilization, just as an increase in the length of intervals between individual applications of up to 2 days, significantly reduces the degree of expression of the sympatho-adrenal reaction in response to repeated immobilizations. This was manifested by a reliably smaller increase in KA and VMK excretion in urine throughout the entire period of stress applications, in the absence of an increase in the level of A discharge under repeated immobilizations, as well as in a more rapid and significant reduction of the increased level of KA excretion by the second half of the stress cycle. Such a change in the dynamics of the KA and VMK excretion in monkeys of the second and third group attests to the earlier development of adaptive changes of the function of the SAS activity and allows us to conclude that preliminary imposition of the stress stimulant or an increase in the length of intervals between applications with a post-application period that is sufficient for the restoration of changed indicators of SAS activity, can be seen as a factor which increases the resistance of this system to ES.

Thus, the adapting effect of preliminary immobilization and the increase of intervals between repeated stress applications was clearly manifest not only in the delimiting of SAS reactivity, but in the optimization of its functioning under conditions of repeated ES. In the experiments conducted, the observed increase of the resistance of the SAS function to the influence of repeated stress stimuli under the influence of a change in schedule of application toward an increase in the duration of individual restorative periods shows that a more pronounced realization of natural self-limiting mechanisms of the extreme activation of SAS requires a certain period of time during which these mechanisms can effectively begin to work.

This aspect is of particular interest, since it attests to the possibility of a directed regulation of the functions of such an important stress-realizing system as SAS, as well as of the processes of adaptation on the whole through a search and selection of the more optimum periods between application of repeated stress stimulants. In analyzing the physiological expediency of the phenomenon of extinguishing the excitement of the stress-realizing systems, which is developed in the process of adaptation to repeated stress applications, Meerson (1984) indicates that the given process occurs when an escape from the stress situations by way of external behavior adaptation is impossible, and presents itself as a unique internal adaptation to unavoidable stress situations, leading in the long run to a decrease in the concentration of KA and gluco-corticoids, operating on the target organs.

The inclusion of these mechanisms decreases the likelihood of stress injuries to internal organs and signifies a transition to a more complete adaptation which, as Meerson feels (1984), is brought about on the basis of the coordination of the inclusion of GAMK--urgic, serotoninergic and other braking systems of the brain, which present on the whole the central braking mechanisms for the suppression of the excitement of the stress-realizing systems. Therefore we can assume that the nature of the SAS response to the influence of stress stimuli from the temporary regimen of their application, revealed in our experiments, reflects a complex change in the central neuro-chemical processes, securing a reduction in the emotional tension under repeated ES, based on the formation of new psychological adjustment to the acceptance of a stress situation as less adverse.

A FUNCTIONAL TEST WITH L-DOPA IN MONKEYS UNDER CONDITIONS OF PHYSIOLOGICAL CALM AND STRESS

At the present time for the analysis of the synthetic abilities of the SAS a functional test with a precursor of DA and a preparation of L-DOPA is widely used. The test with L-DOPA allows us to judge not only the reserve abilities and reactivity in the SAS in a healthy organism, but gives us the opportunity to reveal certain aspects of the disturbance of KA exchange under various pathological processes in man and animals (Anokhina, 1975; Zavodskaya, et al., 1977; Zabrodin, 1982; Mezentseva, 1982; Vasilyev, Chugunov, 1983; Stulaya, et al., 1984b).

The ability of L-DOPA to participate in KA synthesis and thereby lead to the restoration of the mechanisms of sympathetic regulation of the energetic processes has found application in clinical practice for the treatment of ulcer conditions of the stomach and duodenum, as well as in experimental medicine for the purpose of clarifying the role of the disturbance of the adrenergic mediation in the pathogenesis of dystrophic injuries to internal organs, caused by the influence of intensive neurogenic stimulants (Zavodskaya, et al., 1977; Zabrodin, 1982).

For this reason it seemed interesting to explain the synthetic abilities of SAS and the possibility of a directed increase of its functional reserves under the influence of stress stimulants in monkeys in a setting of the introduction of L-DOPA.

The experiments were conducted on 5 intact sexually mature Papio hamadryas males, previously adapted to confinement in individual metabolic cages. The preparation was introduced orally in doses of 0.1 and 0.75 grams per monkey. The choice of the doses of L-DOPA was based on available data on the application of these doses, as the most physiological, for an evaluation of the synthetic abilities of SAS in healthy people and under various forms of pathology (Vasilyev, 1981; Vasilyev, Chugunov, 1983). As a control for the experiments with L-DOPA on intact monkeys and with immobilization in a setting of preliminary introduction of the preparation, data were used which were obtained under analogous experiment conditions with oral introduction of a placebo in an additional group of clinically healthy Papio hamadryous males (N=5).

As our experiments showed, the introduction of L-DOPA in a dose of 0.1g under conditions of psychological calm led, in all monkeys, to the increase of KA excretion in urine (Illustration 6). In this, the excretion of A, NE and DA, relative to the initial amounts, on the average among the 5 animals increased by 1.9 times ($P<0.02$), 2.4 times ($P<0.01$) and 3.2 times ($P<0.0001$), respectively. However, regarding the control (introduction of the placebo), the differences in the excretion of A and NE were statistically unreliable, although the discharge of A and NE

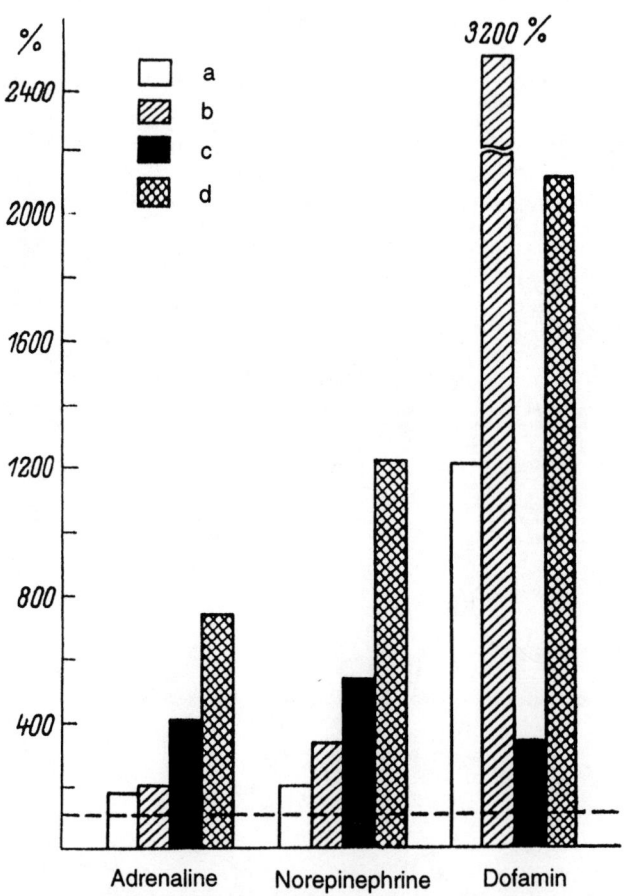

Illustration 6 Test with L-DOPA (0.1g - a, 0.75g - b) in monkeys under conditions of physiological calm and stress.

Y-axis - amounts of daily catecholamine excretion, % of initial figures.
Immobilization - 2-hour: control (b) and against a background L-DOPA (d).

exceeded the corresponding amounts of their excretion in the control experiment by 28% and 61%, respectively. By the following day after the introduction of the preparation a normalization of the amounts of A and NE excretion was noted, whereas the content of DA in the urine for the day remained somewhat elevated, although it did not differ in a statistically reliable manner from the initial level.

The introduction of L-DOPA lead to an insignificant increase in the amount of the coefficient NE/A from 2.32 to 2.92, and a significant reduction in the coefficient NE/DA-from 0.04 to 0.008, as opposed to 0.06 in the control. There was also observed a reduction in the amount of the ratio (A+NE/DA) to 0.01, as opposed to 0.06 in the initial state and 0.09 in the control. A significant increase of DA excretion, along with a significant decrease in the amount of the ratios (NE/DA) and (A+NE/DA) shows that the introduction of L-DOPA in a dose of 0.1g in intact monkeys leads to a predominant increase of DA synthesis.

With the introduction of L-DOPA in a dose of 0.75 g there was observed a sharper and longer elevation of DA excretion in urine (Illustration 6). DA excretion, relative to the initial values, increased by 34.2 times ($P<0.0001$). The discharge of A was elevated 2-fold ($P<0.02$), as with the introduction of the preparation in a dose of 0.1g, the daily excretion of NE increased to (78.86±18.16) n-moles/24 hours, exceeding the baseline values by 3.3 times ($P<0.05$). In this experiment the discharge of A and NA surpassed the corresponding amounts under the introduction of the placebo, by 100% and 153%, respectively, with $P<0.05$. The introduction of a 7.5g dose of L-DOPA led to a reduction of the coefficients NE/DA to 0.004, and of A+NE/DA to 0.006. Consequently, increasing the dose of L-DOPA to 0.75g for 1 animal results - along with the significant increase of A and NE excretion - in an extreme increase of production. Therefore, in order to establish the possibilities for the increase of the functional reserves of the SAS under conditions of ES in monkeys, it is more expedient, in our opinion, to apply L-DOPA in a dose of 0.1g per animal.

During a 2-hour immobilization with the preliminary introduction of L-DOPA, there is observed a more extreme increase in the excretion of KA, in comparison with analogous stress application without the introduction of the preparation (Illustration 6). Furthermore the excretion of A, NE, and DA were greater in comparison with the effect of the immobilization without the introduction of L-DOPA, by 65% ($P<0.05$), by 168% ($P<0.02$) and by 554% ($P<0.0001$), respectively. Under these conditions there was noted a predominance of the activation of the mediator unit of the SAS over the hormonal unit, indicated also by the significant increase in the coefficient NE/A to 4.26, as opposed to 2.63 in the control. In spite of a more severe increase of NE excretion under immobilization with the introduction of L-DOPA, the amount of the ratio (NE/DA) and (A+NE/DA) turned out to be reduced, as the result of a more significant increase in DA excretion, when compared with immobilization without the introduction of the preparation. Thus, the amounts of the coefficients NE/DA and NE+A/DA were equal to 0.03, as opposed to 0.06 and 0.09, respectively, in the control. The expressed changes of the amount of the ratios between KA indicate that under conditions of stress with the preliminary introduction of L-DOPA there is observed an increase in SAS reactivity with a simultaneous increase in its reserve abilities.

Thus, the application of a functional test with L-DOPA in our studies allowed us to reveal the presence of high synthetic abilities of the SAS in monkeys. The data obtained agree with the results of the research of other authors, which points to an increase of KA and DOFA excretion in the urine of healthy people and animals under pharmacological loads by a preparation of L-DOPA (Vasilyev, 1981; Mezentseva, 1982). However, in Papio hamadryas males there was noted a more significant increase in dopamine synthesis. At the same time, in mice the inner-abdominal introduction of L-DOPA in a dose of 50 mg/kg caused a more severe (10-fold) increase of NE excretion (Mezentseva, 1982). The differences revealed can be caused both by the dose and method of introducing a preparation, and by species peculiarities of KA exchange. The predominant increase of dofamin production in monkeys under the introduction of various doses of the preparation can be explained by the rapid decarboxylation of L-DOPA into dofamin with the subsequent (only partial) transformation of DA into NE and furthermore, as shown by Allikmets (1977), the predominance of DA synthesis over NE under the influence of L-DOPA can be related to the saturation of dofamin-hydroxylase.

In monkeys, under conditions of extreme stress, the preliminary introduction of L-DOPA along with the increase of reserve abilities of the SAS leads to higher amounts of KA excretion, which, in our opinion, points to the expediency of using the L-DOPA preparation in order to forestall exhaustion of the KA synthesis under the influence of stress stimulants in a complex with various pharmacological preparations, blocking the secretion and effects of the influence of KA. The prospect for conducting studies in this direction are supported by available data, which point to the increase of the protective effect of the neuro-tropic preparations under conditions of stress in joint application with L-DOPA (Stanishevskaya, Mezentseva, 1977; Mezentseva, 1982).

Chapter 3

GLUCO-CORTICOID HORMONES IN MONKEYS UNDER EMOTIONAL STRESS

Activation of the hypothalamus-hypopituitary-adreno-cortical system in response to the influence of stress stimulants is the most important component of the stereotype non-specific stress reaction of the organism. Gluco-corticoid hormones play an exceptionally important role in the formation of adaptive reactions of the organism of man and animals to the influence of negative factors and to changes in the environment. However, whereas SAS activation has a primary significance in the immediate mobilization of the functions and the energetic resources of the organism, the change in the levels of hormones of the adrenal cortex leads to a longer and more persistent adaptation on the basis of the mobilization of plastic reserve. Along with the mobilization of the plastic and energetic resources of the organism, culminating in the creation of a free amino acid fund, necessary to provide adaptive protein synthesis and increased glyconeogenesis, gluco-corticoids take part in the regulation of VND and in providing immune homeostasis, influence the electrolyte exchange, and have a permissive effect on the influence of catecholamines, etc. (Yudaev, et al., 1977; Korneva, et al., 1978; Korkach, 1979; Viru, 1981; Panin, 1983, and others).

At the present time a great amount of factual material is significantly expanding the concepts of the mechanisms of the development of general adaptive syndrome of Selye (1937, 1960). Many studies have established the leading role of such structures of the brain as the hypothalamus, the hyppocamp, the amygdaloid complex, the septum, the neocortex and its other departments in the regulation of the hypopituitary-adreno-cortical activity (Malyshenko, 1976; Viru, 1978; Aleshin, 1979; Filaretov, 1979; Sapronov, 1980). In this the fundamental role is given to the hypothalamus, as the highest center of the integration of neuro-humoral and neuro-hormonal functions.

Gluco-corticoid secretion occurs in bursts (quanta) with a definite frequency, comprising individual episodes, the distribution of which in the course of a day creates a curve of the C-rhythm of the gluco-corticoid content in the blood; furthermore, episodes of their elevated secretion are preceded by periods of an increased level of AKTG in the blood and of the content of cortico-liberine in the hypothalamus (Krieger, et al., 1971; Hiroshige, Wada, 1974; Ratge, et al., 1982). The daily rhythm of hydro-cortisone and its predecessors in lower monkeys leading a diurnal lifestyle is characterized by a elevated gluco-corticoid content during the morning hours (from 6 to 9 a.m.) and by a subsequent reduction during the evening hours (from 6 to 9 p.m.), revealing a similarity with man in the amounts and the duration of the secretary episodes (Jacoby, Sassin, 1974; Holaday, et al., 1977; Tavadyan, 1981; Taranov, Goncharov, 1981).

It is felt that the genetically predetermined C-rhythm of adreno-cortical activity is produced by an endogenous rhythm driver, whose synchronization with geophysical hours is influenced by such exogenous factors as a change in light, a change in the cycle of the sun--waking, feeding time, social activity, etc. (Graef, Golf, 1979; Kazin, 1982). The adaptive value of the C-rhythm of gluco-corticoids lies in the fact that the elevated secretion of hormones in the blood, noted during various times of day, preceding the beginning of motion activity, lead to the mobilization of the energetic and plastic resources necessary for intense activity of the organism, which ensures the temporal synchronization of inner cellular processes with factors of the external environment (Efremova, 1978; Filaretov, 1979; Kazin, 1982).

It has been established that under the influence of stress stimulants in man and animals there occurs a change in the daily rhythms of gluco-corticoid secretion. As shown in experiments on rats, the nature of the C-rhythm change of cortico-sterone under emotional-muscular stress (swimming) depends on the phase of the daily and seasonal variations of adreno-cortical activity, with which the influence of the stress stimulant coincides (Efremova, 1978; Kazin, 1982). Thus, frequently repeated physical loads, coinciding with the period of maximum motion activity (evening hours), assists the synchronization of the endogenous rhythm of adreno-cortical activity, shifting somewhat the acro-phase into the pre-work period, while muscular loads in the morning hours lead to a reduction in the amplitude of the basal C-rhythm of gluco-corticoid secretion in spring months, and its disappearance during the fall.

In healthy people, during intensive neuro-emotional work during the day there is noted an increase in adreno-cortical activity and an insignificant shift of the acrophase of the daily rhythm of gluco-corticoid secretion, while similar work during the night results in a disturbance of the daily rhythms of GGAKS activity, a restoration of which occurs only by the second day of rest (Belova, Vasilyev, 1974). Changes which have been studied in the structure of the C-rhythm of gluco-corticoid secretion in man and animals also arise with the prolonged influence of a complex of negative technogenetic factors, and are accompanied by the development of functional changes of the somato-visceral systems (Kazin, 1982).

A noticeable reduction in the amplitude of daily rhythms of gluco-corticoid secretion in monkeys--male rhesus macaques--was revealed under clinostatic hypokinesia (Tavadyan, 1981). A complete disturbance of the C-rhythm of the hydrocortisone content in the blood was noted in sexually mature Papio hamadryas males under a 2-hour immobilization (Taranov, 1981).

Available data on the study of the interrelationships of C-rhythms on the secretion of gluco-corticoids and the GGAKS reaction to stress indicate a close mutual relation of these processes (Efremova, 1978; Filaretov, 1979; Tavadyan, 1981; Kazin, 1982).

A study of the nature of the hormonal reaction of the adrenal cortex in sexually mature Papio hamadryas males under severe ES in morning and evening hours, conducted in our laboratory by M.G. Tsulaya (1985), allowed us to reveal clear differences in the degree of expression of the adreno-cortical response depending on the daily rhythm of the secretion of gluco-corticoids. As Illustration 7 shows, during immobilization in the morning hours (10:00-12:00), a significant increase in the level of gluco-corticoids in the blood was noted in monkeys. The maximum increase of hydrocortisone in the blood was observed 2 hours after the beginning of the stress application, and the concentration of the hormone amounted on the average for 5 monkeys to (1568±64) n-moles/l, which exceeded the initial level by 96% ($P<0.0001$). The elevated hydrocortisone content remained even after 6 hours, reliably not differing from its maximum level, noted 2 hours after the beginning of the application. The restoration of hydrocortisone concentration in the blood to the initial values occurred the following day after the immobilization. A similar dynamic was revealed in 11-desoxyhydrocortisone, the amount of which 2 hours after the beginning of immobilization exceeded the initial level by 1.8 times, with $P<0.01$. The content of hydrocortisone 2 hours after the beginning of the stress application amounted to (103.9±16.4) n-moles/l, which was 2 times greater than ($P<0.02$) its initial value. The normalization of the amounts of the content of 11-desoxyhydrocortisone and corticosterone were also noted 24 hours after the beginning of the influence of the stress stimulant. Under conditions of severe ES a highly positive correlation was revealed in the dynamics of hydrocortisone with 11-desoxyhydrocortisone ($r=0.99$, $P<0.01$) and corticosterone ($r=0.89$, $P<0.01$). The size of the ratio of hydrocortisone/corticosterone in monkeys 2 hours after the beginning of immobilization was 15, as opposed to 16 in the initial state, and it increased 6 hours later to 19. In 24 hours, some reduction was noted in the ratio hydrocortisone/corticosterone, with normalization by the

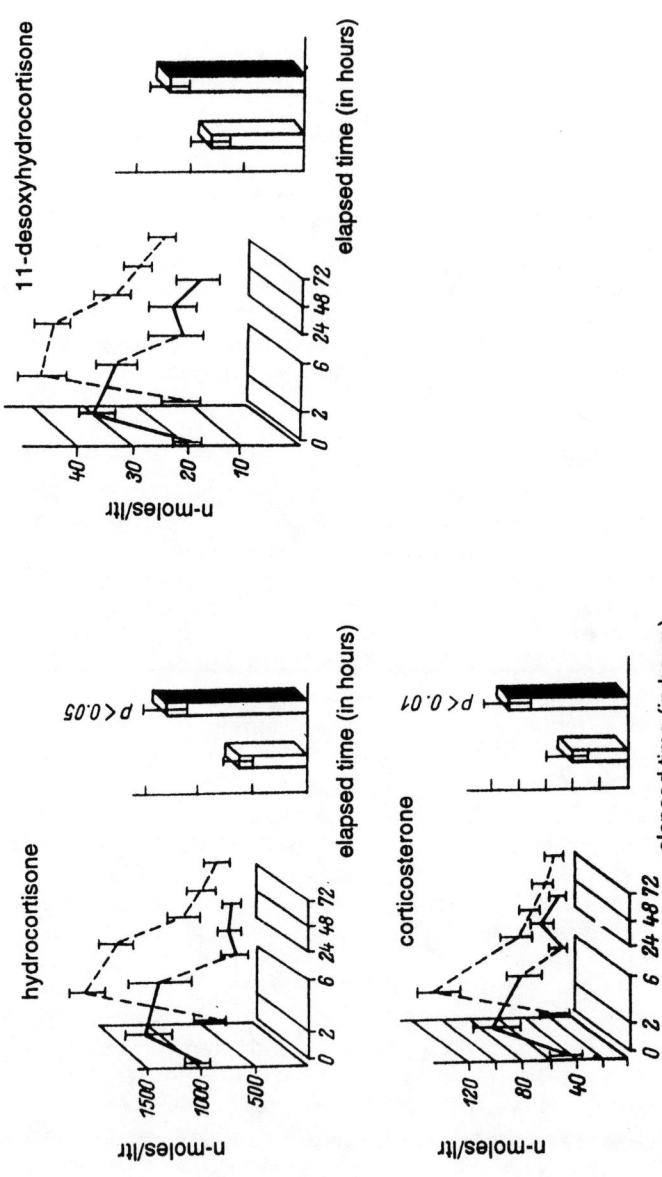

Illustration 7. Changes in glucocorticoid concentration in the blood of monkeys under stress in morning (solid line) and evening (dotted line) hours. Mathematical averages with reliable intervals.

Y-axis - hormone concentration, n-moles/ltr; X-axis - time of blood draw: 0 - before immobilization, 2, 6, 24, 48, 72 - time elapsed since the start of immobilization, in hours; clear columns - amplitude of steroid concentration increase under stress in the morning (10:00-12:00); darkened columns - amplitude of hormone concentration increase in the evening (22:00-24:00) third day. The amounts of the ratio of hydrocortisone/corticosterone revealed 6 hours after the beginning of immobilization evidently point to a relative predominance, under conditions of severe ES, of the increase of hydrocortisone biosynthesis over corticosterone.

Similar data were obtained previously in studies by Goncharov, et al. (1977a, 1978a) and Taranov (1981), which showed that under the influence of a 2-hour immobilization in sexually mature Papio hamadryas males, along with an increase in the level of hydrocortisone and of 11-desoxyhydrocortisone in the blood, there occurs a significant elevation of the concentration of pregnenolone and 17-hydroxilized predecessors of hydrocortisone: 17-oxypregnenolone and 17-oxyprogesterone.

More pronounced changes of adreno-cortical activity were observed in monkeys during immobilization in the evening hours (10:00-12:00 p.m.), i.e., during the phase of the daily reduction of gluco-corticoid production (Illustration 7). The amplitude of the increase of gluco-corticoids in animals of this group significantly exceeded the amount of the increase of these hormones under stress during morning hours. Thus, the amount of the amplitude of the increase of hydrocortisone under stress in the evening hours amounted to, on the average, (1209 129) n-moles/l, which exceeded the amplitude of the increase of the hormone during immobilization in the morning hours by 1.6 times ($P<0.02$).

The amplitude of the increase in the level of 11-desoxyhydro-cortisone and corticosterone under stress during the evening amounted to (24.7±1.7) n-moles/l and (90.3±12.9) n-moles/l, respectively, and exceeded the amount of the amplitudes of the increase of these steroids under immobilization during the morning hours by 1.5 ($P<0.05$) and 1.8 times ($P<0.05$), respectively. At the same time the absolute amounts of the concentration of these hormones 2 hours after the beginning of the immobilization both in the morning and in the evening hours had virtually no statistically reliable differences. The larger amplitude of the increase of gluco-corticoids under stress during the evening was caused by a low basal level, which corresponded to the daily rhythm of the secretion of these hormones. The restoration of the amounts of the gluco-corticoid content in monkeys of the second group to the initial values was observed only by the second and third day of the after-effect.

Under stress during the evening hours there were noted also more pronounced changes in the size of the ratio hydrocortisone/corticosterone. It should be noted that among intact monkeys the amount of this ratio during the evening increases to 24, as opposed to 16 during the morning hours. 2 hours after the beginning of the immobilization during the evening the amount of the ratio hydrocortisone/corticosterone was reduced to 15, and then sharply increased, reaching a value equal to 32, 6 hours after the beginning of the application. The established differences in the amounts of the ratios hydrocortisone/corticosterone, both in the initial state and under severe ES in morning and evening hours, can be caused by daily characteristics of the biosynthesis of gluco-corticoids and by the increase in sensitivity of the adrenal cortex to AKTG.

The results of the study conducted agree closely with data of Markel, et al. (1981), which indicate that between the basal rhythm of the concentration of cortico-steroids and the response reaction to stress applications there is a pronounced reciprocity: With a high basal level of hormones in the blood, the reaction to stress is less than with a low basal level. In the opinion of this author, a high initial level of hormone according to the mechanism of negative reaction can lower the sensitivity of the hypopituitary-adrenal system to stress applications and thus, the rhythm of the stress reactivity is like a derivative of the rhythm of the basal level of the functioning of GGAKS.

Numerous data obtained in experimental and clinical research show that the change in the functional status of GGAKS in various animals and in man takes place under the influence of a broad spectrum of stress stimulants, including psychogenic and physical loads, the influence of hypokinesia, hypothermia and cooling, the geomagnetic field of the earth, noise and other technogenetic factors, hypoxia, anesthesia, chemical and various medicinal preparation, etc., and also in burns, pain and traumatizing influences, operative interferences and illnesses of the widest variety (Goncharov, 1971; Vorontsov, 1972; Belova, Vasilyev, 1974; Mikulai, et al., 1974; Bankova, et al., 1976; Tang, Phillips, 1977; Chamove, Bowman, 1978; Davydova, 1978; Kulagin, Davydova, 1978; LeMevel, et al., 1979; Shurygin, 1980; Gubachev, Stabrovsky, 1981; Kazin, 1982; Robu, 1982; Fenske, et al., 1982; Sokolova, Belova, 1983; Uchakina, 1983; Axelrod, Reisine, 1984; Kigranyan, 1985).

Especially informative is the quantitative and qualitative evaluation of hormonal reactions to various stress applications, received from experiments on monkeys as a result of their great similarity to man in the spectrum of produced steroid hormones, biorhythms of gluco-corticoid secretion, their blood content level, processes of metabolism, and mechanisms of the regulation of GGAKS under stress (Goncharov, 1971; Yudaev, et al., 1976; Goncharov, et al., 1977a, 1978b; Butnev, 1980; Tavadyan, 1981; Taranov, 1981; Chirkov, 1984).

Studies conducted under the direction of Professor N.P. Goncharov on the secretion of steroids in monkeys have allowed us to study in depth the spectrum of secreted hormones, and to clarify the particular participation of the adrenals and the gonads in the formation of the steroid pool, circulating in the blood. It has been shown that the adrenals in sexually mature male Papio hamadryas secrete hydrocortisone, 11-desoxyhydrocortisone and corticosterone in larger amounts (68%, 10.8%, 7.1%, respectively). In the adrenals 17-oxyprogesterone, 17-oxypregnenolone, pregnenolone and dehydroepiandrosterone are produced in significant amounts, while progesterone is found in trace amounts. On the basis of the data obtained, the authors draw their conclusion about the realization of hormone synthesis in the adrenals of Papio hamadryus primarily along the lines: pregnenolone to 17-oxypregnenolone. Closest to man in the amount of the relation of hydrocortisone to corticosterone are Papio hamadryas, in whom this indicator reaches 10, while in rhesus macaques it approaches 20. Based on the results of a comparative study of the metabolism of the main C-subscript 21-steroids and certain C-subscript 19-steroids a single purpose nature of the metabolism of gluco-corticoids in lower monkeys and in man was revealed (Goncharov, et al., 1977a).

Interesting data on the nature of the hormonal response of the adrenal cortex in monkeys (Papio hamadryas, rhesus macaque) under stress situations were obtained by Goncharov (1971), and show that even the procedure of submitting monkeys to experiments leads to an increase in the gluco-corticoid content in the blood. Observing standard conditions for conducting the experiment, this reaction quickly was extinguished, and by the third experiment no increase in the level of gluco-corticoids in the blood was noted. Physical strain (running in a wheel) in female Papio hamadryas elicited, within 10 minutes of the beginning of the application, an increase in the concentration of steroids in the blood, from 48.0±1.9 to 77.0±9.7 mg %. This author emphasizes that in the given experiment the activating influence on GGAKS, besides the muscle load, produced a neuro-emotional strain. In response to isolation of male Papio hamadryas and rhesus macaque from a group of females for 90 to 120 minutes, among the males the concentration of 17-OKS exceeded the initial concentration by 50-73%. In the research of Vorontsov (1972) on the study of the dynamics of the discharge of cortical steroids in urine of male rhesus macaques during a 3-hour immobilization, there was a 9-fold increase in the level of hydrocortisone excretion. A characteristic of the metabolic change of gluco-corticoids in these experiments was an increase in the share of unchanged steroids and a reduction in the percent of their tetrahydro-producing and conjugated forms. The normalization of the metabolism and of the excretion of gluco-corticoids occurred in the following 2 days.

The placement of monkeys in individual cages and primatological chairs can lead to an increase in the level of gluco-corticoids in the blood (Mason, 1972; Matsubayshi, 1975), as can lengthy immobilization--hypokinesia (Tavadyan, 1981), anesthesia and surgical interference (Goncharov, 1971), density (Sassenrath, et al., 1969; Sassenrath, 1972), isolation of the male from the female (Goncharov, 1971), removal of the young from the female (Rose, et al., 1968), placement with a more aggressive animal of the same species (Chamove, Bowman, 1978), as well as the influence of other factors of setting and experimental procedures (Brown, et al., 1968; Goncharov, 1971; Mason, 1972; Faucheux, et al., 1976).

A significant increase in the level of cortico-steroids in the blood is observed in monkeys during experiment neurosis, infectious diseases, acclimatization and other pathological states (Goncharov, 1971; Goncharov, et al., 1977a). Furthermore, a great similarity is revealed in the dynamics of the change of gluco-corticoid content in the blood and in the degree of expression of the disturbance of the steroid balance in monkeys with the nature of the response reaction of the adrenal cortex in man, which accompanied burns, surgical interferences, myocardial infarction, serious traumas, etc. (Kulagin, Davydova, 1978; Aono, et al., 1972a), which is a significant fact on which to base the choice of monkeys as the most adequate research object for the further study of the endocrine function of the adrenal cortex under stress.

At the present time we may consider it established that the increase of hypothalamus-hypopituitary-adrenal activity in the form of a stress reaction (counter-shock reaction) is characterized by its cyclicity, which is manifested in the consecutive replacement of the phase of initial increase of the activity of the system by a phase of sub-normal activity with subsequent repeated increase (Viru, 1981). Furthermore, the multi-phase nature of the response to the influence of stress stimuli is manifested not only in the reaction of the adrenal cortex, but also encompasses the activity of other units of GGAKS: the hypothalamus, and the hypopituitary. The increase of gluco-corticoids, the increase of the level of AKTG and the increase of the activity of cortico-liberine in response to the influence of various stress factors (laparotomy, ether anesthesia, adrenalectomy, immobilization, the influence of acoustic and neurogenic stimulants) has a clearly pronounced 2-phase nature (Hiroshige, et al., 1971; Daloman, et al., 1972; Cook, et al., 1973; Sato, et al., 1975; Viru, 1981). To explain the phasic nature of the adreno-cortical reaction under stress, Viru (1981) draws on the mechanisms of reaction, the phenomenon of the periodicity of the secretion of gluco-corticoids, and also the cyclical nature of their influence on a cellular level.

Most important in the regulation of gluco-corticoid secretion are the mechanisms of reaction between the level of gluco-corticoid content and AKTG in the blood and the activity of the hypothalamus-pituitary-adrenocortical system. The reaction mechanisms may be divided both according to the time of their development--quick and delayed--and according to the level of gluco-corticoid and AKTG influence--lengthy, short and ultra-short (KPF-pituitary)(Sayers, Sayers, 1947; Hodges,

Vernikos, 1959; Dallman, et al., 1972; Kendall, et al., 1972; Riegle, 1973; Jones, Hillhouse, 1976; Hillhouse, Jones, 1976; Jones, et al., 1977; Filaretov, 1979; Bogdanov, et al., 1982). It has been shown that quick reaction mechanisms operate under a determined gradient of the increase in the amount of gluco-corticoids in the blood, while the engagement of delayed reaction mechanisms depends on the level of these hormones in the blood (Jones, Hillhouse, 1976; Jones, et al, 1977; Filaretov, 1979). It is assumed that both mechanisms can close on the level of the pituitary. The hypothalamus level of the realization of the mechanisms of quick and deferred reaction is shown in experiments involving the destruction and the electrical irritation of various parts of the hypothalamus, leading both to an increase, and a slowdown in the adreno-cortical reaction to stress (Filaretov, 1979; Brodish, 1979).

It has been established that the introduction of exogenous hormones and the increase in gluco-corticoid secretion in response to stress limit the degree and the length of the GGAKF reaction (Dallman, et al., 1972; Sakakura, et al., 1976; Bogdanov, et al., 1982). Adding to the well known contradiction between the reaction mechanisms based on the level of cortico-steroids, there are data which show the possibility of a new increase of gluco-corticoids and AKTG in a setting of their elevated content in the blood, caused by the previous stress application (Sakellaris, Vernikos-Danellis, 1975; LeMevel, et al., 1979). Thus, under repeated psychogenic stress in rats LeMevel, et al. (1979) discovered an increase in cortico-sterone and AKTG in the blood independently of application time of additional stress stimulus. At the same time, in the experiments of Sakakura, et al. (1976) repeated stress application, produced 5 minutes after a 2-minute immobilization, did not lead to a second activation of GGAKF. Engaging the reaction mechanism under these conditions occurred as a result of a more elevated secretion rate of cortico-sterone, which under the influence of immobilization increased to 9.25 mkg per 1 min, while under conditions of psychogenic stress the rate of secretion of cortico-sterone amounted only to 1.4 mkg per 1 min.

At the same time, according to Viru (1981), a negative reaction cannot be seen as the main mechanism for GGAKF activation in the initial period of stress influence, which is indicated by the absence of the initial pre-stress reduction of gluco-corticoids in the blood, by a high release rate of KRF and AKTG, by the possibility of a new increase in the peak of hydrocortisone against a background of its high content in the blood, and by a response reaction of AKTG to stress in adrenalectomized animals. It cannot be seen as a trigger mechanism and the influence of an increased burst of adrenaline from the cortico-matter of the adrenals both on the function of the adeno-pituitary, and on the central adreno-reactive structures.

The latter stems from works which show a reduction of the response adreno-cortical reaction to the repeated introduction of adrenaline, while maintaining a response to other stimuli, and also the possibility of GGAKF activation against a background of the demedullation and the blockade of A-adreno receptors (Selye, 1960; Redgate, 1970; Hiroshige, 1973; Shalyapina, 1976; Viru, 1978, 1981; Filaretov, 1979; Sapronov, 1980; Manukhin, et al., 1981). As Viru (1981) feels, neither reaction mechanisms nor an elevated adrenaline secretion from the cerebral matter of the adrenals can compete with the neurogenic mechanism, which is a trigger mechanism in the activation of GGAKF in the initial stage of the application of stress irritants.

Studying the complex nature of the interaction of the parameters of intensive stress irritants, the initial functional activity of GGAKF and the engagement of the braking mechanism, limiting the development of stress reaction, Bogdanov, et al. (1982) established that increasingly stronger electrical pain irritant in rabbits causes a gradual increase in the response of the cortex of the adrenals, and furthermore it has an optimum force beyond which a gradual reduction in the adreno-cortical reaction is noted. As experiments show, a slowdown in the stress reaction of GGAKF in response to immobilization depends on the dose of hydro-cortisone introduced: the greater the initial level of cortico-steroids in the blood, created by the exogenous hormone, the greater the subsequent suppression of gluco-corticoid secretion under stress. The importance of the quantitative measure of the irritant and the degree of its biological activity in the response activation of the cortex of the adrenals and the overall adaptational reaction of the organism is also shown by the research of Garkavi, et al. (1977). A pronounced dependence of the GGAKF reaction to the duration of stress was determined in the experiments of Robu (1982), who showed that along with the preservation or destruction of actions and reactions according to a vertical level of subordination of the system of hypothalamus-pituitary-adrenal cortex, the content of free hydro-cortisone in the blood under stress is influenced by the state of the peripheral tissue-receptor sub-system.

It is felt that the reduction in the adreno-cortical activity under stress, which comes after the initial increase in the gluco-corticoid level in the blood, forestalls the development of course functional and morphological disturbances in the GGAKF. Furthermore the reaction mechanism operates, as a rule, until the appearance of signs of exhaustion of the adrenal glands and the hypothalamus, thereby providing for the preservation of hormonal resources and limiting the influence of steroid hormones on a tissue and cellular level (Cook, et al, 1973; Viru, 1981). Following the reduction in the adreno-cortical activity (sub-normal phase) there comes a period (phase) of the secondary activation of the GGAKF, furthermore the change in these phases and their duration depend on many factors, including the strength and duration of the stress irritants.

Taking into account the fact that the quantitative relationship between the size of the stress irritant and such a braking stimulant as an increase in the level of endogenous gluco-corticoids in the blood has been insufficiently studied, we established as our task the clarification of the reserve possibilities of the adrenal cortex in sexually mature monkeys under conditions of extremely intensive and lengthy activity of the stress irritant--a 10-hour immobilization. The data received in this experiment show that under the influence of lengthy ES in monkeys there occurs a sharp increase in the level of hydro-cortisone in the blood, which is maintained throughout the entire stress period (Illustration 8). Thus, 2 hours after the beginning of immobilization the concentration of this hormone amounted to (2068 144) n-moles/l and continued to increase, reaching maximum sizes of (2722 255) 7 hours after the beginning of the application. The hydro-cortisone content in the blood in this period exceeded its initial values by 119% ($P<0.002$). The high level of the hormone was maintained to the end of the stress application, and 2 hours after its termination the concentration of hydrocortisone in the blood not only was not reduced, but corresponded to the maximum values registered under conditions of stress. 24 hours after the beginning of immobilization the content of hydrocortisone in the blood did not differ from the initial values and after its insignificant but reliable reduction ($P<0.05$) after 48 hours the concentration of the hormone had again returned to the initial level.

The results of the research show that a 2-fold increase in hydro-cortisone concentration in the blood during the period of the activity of the stress irritant does not lead to a slow-down in adreno-cortical activity, since the content of the hormone in the blood remained high throughout the entire period of immobilization. A reduction in the level of hydro-cortisone 48 hours after the beginning of immobilization with its subsequent increase to normal levels by 72-96 hours to a certain extent reflects the phasic nature of the reaction of GGAKF, however we did not discover a phase of secondary activation of the system in the periods of time which we studied. The data obtained point to a significant difference in the dynamics of the change of adreno-cortical activity in monkeys under stress in comparison with small laboratory animals, in whom the phase of sub-normal activity during stress irritation of comparable intensity begins in most cases significantly earlier (Dallman, et al., 1972; Dallman, Jones, 1973; Viru, 1981; Robu, 1982).

At the same time, considering the intensity of this stress effect applied in our experiments, leading, as shown in the second chapter, to a sharp burst of KA and a serious general somatic condition of the monkeys, one can assume that the preservation of the high secretory activity of the adrenal cortex under these conditions reflects a disturbance in the reaction mechanism according to the level of gluco-corticoids in the regulation of the GGAKF, and in particular in limiting the length of its activation. Proof of this is the large number of works in which it is convincingly shown that under a lengthy activity of stress irritants, and also under serious pathological processes there occurs an error in the self-regulating mechanisms of the GGAKF and the exhaustion of the function of the adrenal cortex (Selye, 1937, 1960; Goncharov, 1971; Kulagin, Davydova, 1978; Shurygin, 1980; Viru, 1981).

The more complex nature of the functional restructuring of the adreno-cortical activity is found under conditions of chronic and repeated activity of stress irritants. What's more the gluco-corticoid content in the blood, reflecting more objectively the secretory activity of the adrenal cortex, is influenced by the condition of the central nerve structures, which ensure the engaging of the self-regulating mechanisms of GGAKF at various levels of the organization of this system, the processes of the interaction of gluco-corticoids with proteins of plasma and with receptor steroid-sensitive structures of tissues, the rate of their metabolic transformations, and also the state of the functional activity of other neuro-endocrine complexes--the hypothalamus-pituitary-thyroid complex, the hypothalamus-pituitary-gonad complex, etc..

It should be noted that available data dealing with the change in the size of the adreno-cortical response under the influence of repeated stress irritants, are very conflicting. A great amount of factual material attests to the fact that under the influence of repeated stress irritants there occurs a decrease in the level of activation of GGAKF. Thus, repeated 2-hour immobilizations, conducted twice a day for a period of 20 days, in mice lead to a reduction in the size of the increase of cortico-sterone in the blood by the final applications, as well as a decrease in the adreno-cortical response to additional ethereal stress (Riegle, 1973). However, there was not observed a complete suppression of adreno-cortical reaction in these experiments, and furthermore the reduction in the increase of the level of cortico-sterone in the blood was not related to a disturbance in the function of the adrenal cortex, since the test with AKTG elicited the preservation of their high secretory activity. Analogous data were obtained in the research of Mikulai, et al. (1974), in which, in spite of a distinct progressive reduction in the adreno-cortical response in rats to repeated 150 immobilizations, the content of cortico-sterone in the blood after the 42nd stress application was statistically reliably higher than the initial content. Furthermore the suppression of the reaction of the adrenal cortex was also not caused by their exhaustion, since in response to stimulation by exogenous AKTG there was noted a significant increase in the level of cortico-sterone in the blood. Bankova, et al. (1976) note that in rats, in

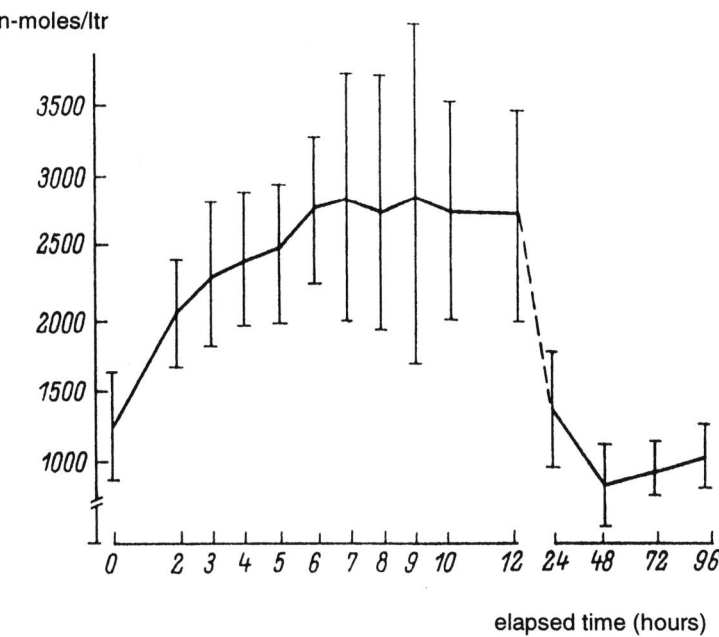

Illustration 8. Dynamics of hydrocortisone concentration in the blood of monkeys under prolonged emotional stress. Mathematical averages with reliable intervals.

Y-axis - concentration of hydrocortisone, n-moles/ltr; X-axis - time of blood draw: O - before immobilization, 2,3,4...96 - time elapsed since the start of immobilization, in hours.

response to an increased psychogenic stimulation after increased production of cortico-sterone, observed in the first 24-hour period, there occurred a gradual reduction of its production and by the 15th day there had taken place a decrease in the mass of the adrenals and a reduction in the biosynthesis of cortico-steroids.

It is interesting to note that during repeated immobilizations of rats both under conditions of free maintenance and in a setting of hypokinesia an increase of cortico-sterone in the blood and morphological changes in the adrenals were less pronounced in comparison with severe stress (Savina, et al., 1980). The immobilization of rats during the course of 60 minutes, conducted during a period of persistent reduction of cortico-sterone, caused by a 12-hour unbroken acoustic stress, lead to an insignificant increase in the rate of secretion of cortico-sterone by comparison with the response to immobilization without a preliminary stress application (Henkin, Knigge, 1963). The suppression of the secretion of cortico-sterone under conditions of acoustic stress and the reduction of the adreno-cortical response to immobilization in these experiments was not related to the exhaustion of the adeno-pituitary and the adrenal cortex, since the AKTG content in the pituitary was 2 times greater than in the unstressed rats, while the introduction of exogenous AKTG caused a significant increase of cortico-sterone in the blood.

Similar data were obtained in male rhesus macaques during a lengthy hypokinesia, which lead to a severe initial increase of hydrocortisone and its predecessors in the blood, which was replaced by a normalization, and later a reduction of their concentration (Tavadyan, 1981). A test with AKTG, conducted under these conditions during a period of a decrease in gluco-corticoid concentration in the blood, revealed an equal and even somewhat elevated response of the adrenal cortex in comparison with the control.

A pronounced suppression of the adreno-cortical response in monkeys was revealed during a lengthy unbroken application of psychogenic irritants (Holst, 1972), as well as during repeated series of flight reaction, which, along with the preservation of the manifestations of behavioral components of emotional excitement (fear) indicates the relativity of the stress criteria according to the amount of the increased level of gluco-corticoids in the blood (Natelson, et al., 1976).

However, the research of other authors obtained data which indicate an increase in the adreno-cortical response under conditions of repeated and chronic stress. In the experiments of Lemevel, et al. (1979) repeated psychogenic stress led to a sharp increase in the level of AKTG and cortico-sterone, exceeding the response to the first application by 2-3 times.

An increase in the response of the GGAKF is indicated by the experiments of Amiragova, et al. (1979), in which under repeated stress applications (a combination of immobilization with an irritant electric shock), produced in dogs, one month after the end of the first series of stress applications there was noted a more pronounced adreno-cortical reaction, which was manifest in the elevation of the base line level of gluco-corticoids and in the large increase of their concentration in the blood. The application of repeated stress irritants (electric shock, laparotomy) 1, 4, 8, 16 and 24 hours after the beginning of the first application led to a more significant or equal increase in cortico-sterone in the blood of rats, and furthermore the application of additional less intense stress stimulai (injection), applied 3 hours after a 90 minute immobilization, elicited a response reaction of cortico-sterone, while 3 hours after the introduction of exogenous cortico-sterone or AKTG in doses corresponding to the increase of the concentration of these hormones in the blood during immobilization, the response to the injection was decreased or not present (Dallman, Jones, 1973). Based on these data the authors come to the conclusion that stress elicits an increase in sensitivity of the CNS to repeated stress irritants, which is equal to the suppressing activity of gluco-corticoids on the corresponding steroid sensitive structures of the brain.

Under conditions of chronic psychogenic stress in monkeys, elicited by a change in their conditions of captivity there was also observed a lengthy increase in the level of hydro-cortisone in the blood and an increase in the response of the adrenal cortex to the introduction of AKTG by 3-10 times in comparison with the control (Sassenrath, 1972; Chamove, Bowman, 1978).

In our opinion, the contradictory data about the change in the adreno-cortical activity to repeated stress applications is caused by the application of various stress irritants, differing in quality, strength and length of application, the frequency and duration of their application, the length of intervals between application of repeated stress, peculiarities of the function of steroid-producing glands in various species of laboratory animal, and is also caused by possible differences in the initial state of the gland, related to seasonal and diurnal rhythms of secretory activity, the use of various methods to determine steroid content, and also differences in method approaches and the interpretation of the results of the research. In addition, at the present time changes in the summary level of gluco-corticoids circulating in the blood have been more thoroughly studied, while the dynamics of the physiologically more active fractions of hormones, i.e., free hydro-cortisone and its predecessors, the determination of which has a primary significance for the analysis of the functional state of the gland, has been researched under conditions of stress, particularly repeated stress, obviously insufficiently.

Therefore for a clarification of the question being examined we conducted a study of the nature of the response adreno-cortical reactions according to the dynamics of the change of the content of hydro-cortisone and its predecessors--11-desoxyhydrocortisone, 17-oxyprogesterone and 17-oxypregnenolone in the blood in 3 experimental groups of monkeys under repeated immobilization with various lengths of intervals between individual applications.

The cycle of daily 2-hour immobilizations led in monkeys to a sharp increase in the level of gluco-corticoids in the blood in response to each repeated stress application (Illustration 9,a, 10,a, 11,a). The concentration of hydro-cortisone during the first immobilization increased within 2 hours of the beginning of the application by 1.4 times and continued to increase, amounting to (1740±101) n-moles/l 6 hours later, which exceeded the initial level by 80%. In 24 hours the content of the hormone had returned to the initial values. Repeated

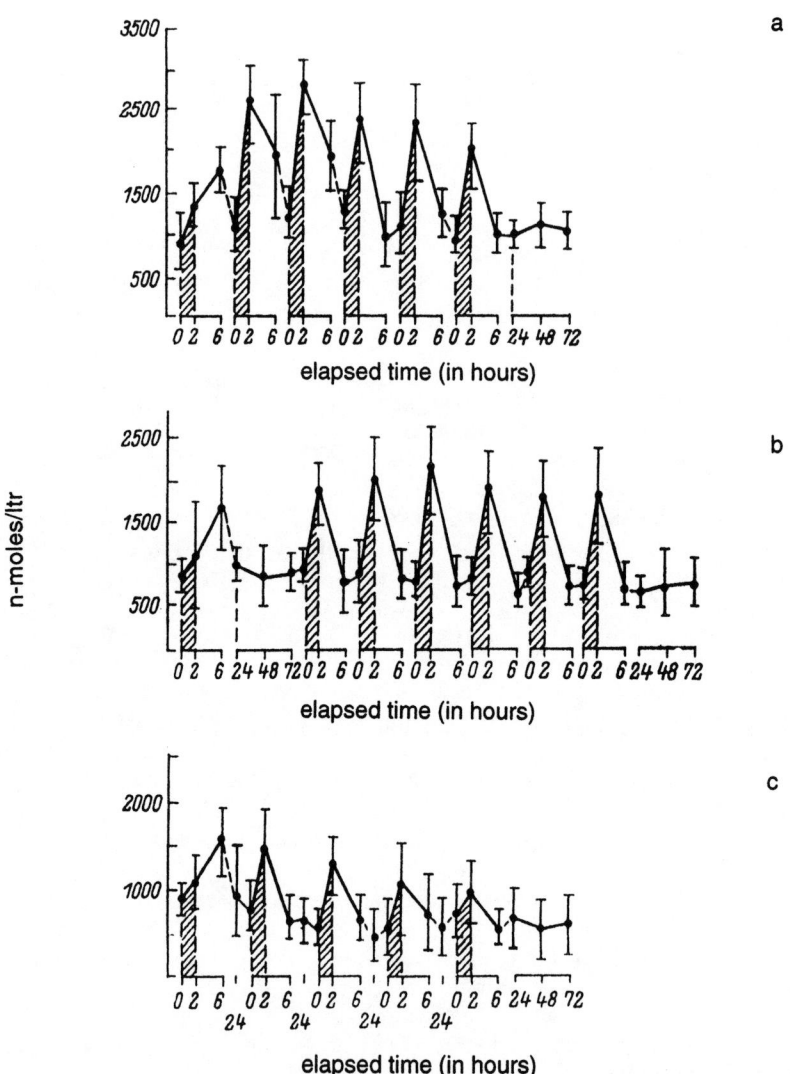

Illustration 9. Nature of the change in hydrocortisone concentration in the blood of monkeys under repeated emotional stress. Mathematical averages with reliable intervals.

Y-axis - concentration of the steroid, n-moles/ltr; X-axis - - time of blood draw: O - before immobilization, 2,3,4...96 - time elapsed since the start of immobilization, in hours; a - cycle of daily 2-hour immobilizations; b - analogous cycle of stress applications with preliminary (3 hours prior) immobilizations; c - cycle of repeated 2-hour immobilizations, separated by 2, 24-hour intervals; striped sections - 2-hour immobilizations.

Gluco-Corticoid Hormones in Monkeys   67

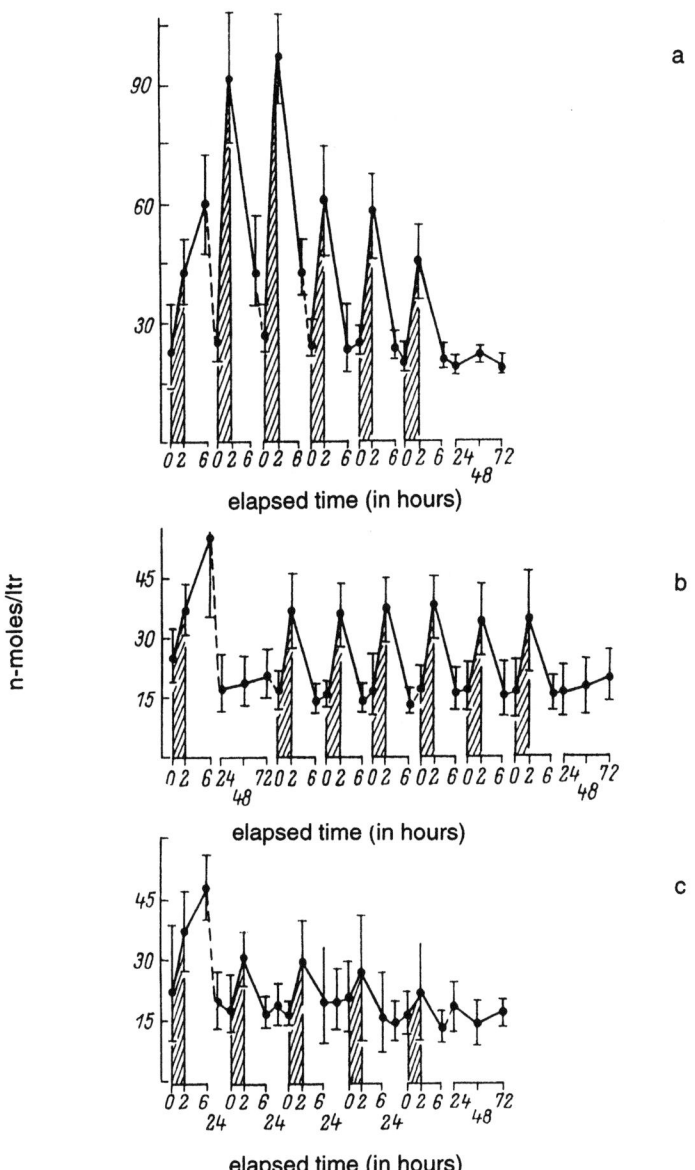

Illustration 10. Nature of the change in 11-desoxyhydrocortisone in the blood of monkeys under repeated emotional stress. Mathematical averages with reliable intervals.

Key: same as Illustration 9.

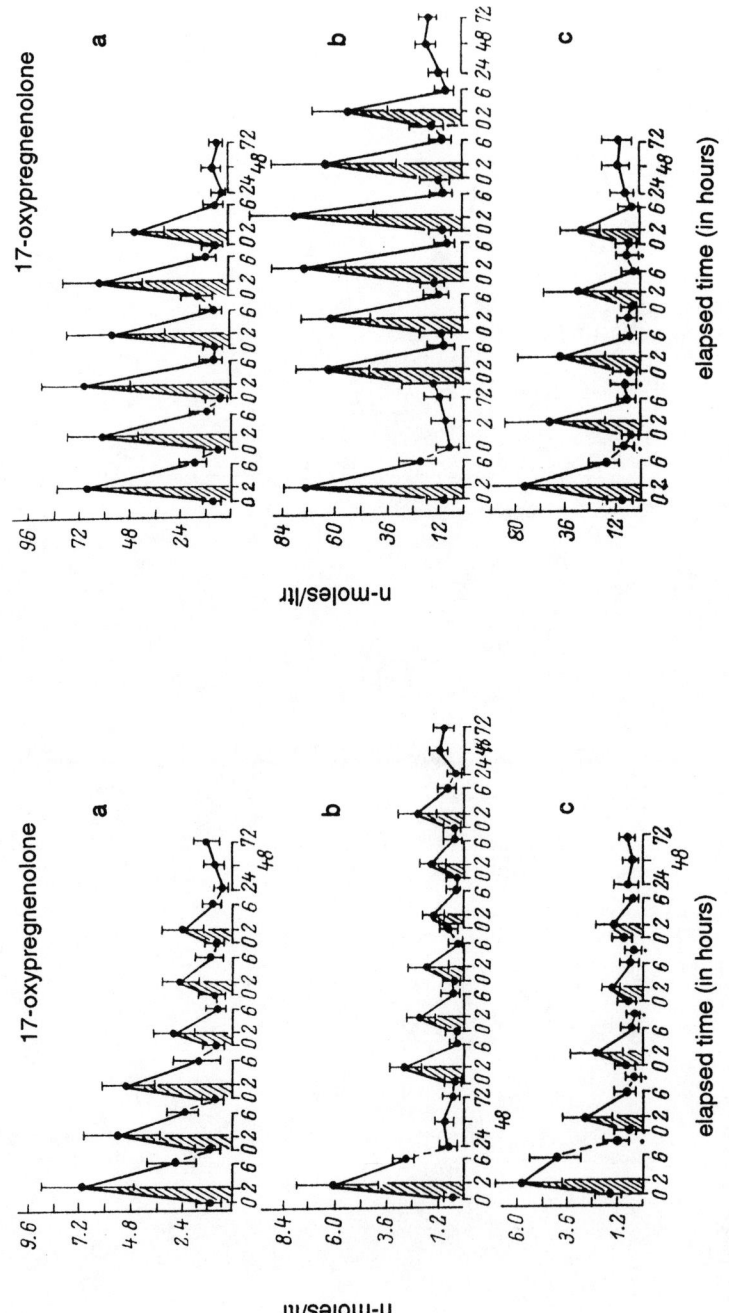

Illustration 11. Nature of the change in the concentration of 17-hydroxylized precursors of hydrocortisone in the blood of monkeys under repeated emotional stress. Mathematical averages with reliable intervals.

Key: same as illus. 9.

immobilizations led to a significantly large increase in the level of hydrocortisone and simultaneously changed the nature of its response reaction: Maximum amounts of hormone content were registered 2 hours after the beginning of each repeated stress application, while 6 hours later their reduction was observed (Illustration 9,a). The amounts of the increase of hydro-cortisone at the second and third immobilizations were maximum and amounted to (2572±180) n-moles/l and (2782±139) n-moles/l, respectively, exceeding the initial level by 135 and 145 percent ($P<0.0001$). In the following immobilizations there was noted a certain reduction in the peak amounts of the hormone, and the increase in its concentration in the fourth, fifth, and sixth immobilizations amounted to 109, 121 and 103%, respectively. A complete restoration of hydro-cortisone content at 6 hour intervals was noted only at the final stress application, while at the second and third immobilization the concentration of the hormone during the period remained sufficiently high and exceeded the initial level by 63 and 55% ($P<0.02$).

In monkeys of the second group, subjected to an analogous cycle of stress applications 3 days after a 2-hour immobilization, there was noted a definite similarity in the response dynamic of the content of hydro-cortisone in the blood with the nature of the adreno-cortical reaction in animals of the first group (Illustration 9,a,b). This was manifest in the subsequent replacement of the increase (in 2 hours) of the hydro-cortisone level by the restoration (in 6 hours) of its content to the initial values in response to each immobilization. However, in animals of the second group throughout the entire 6-hour cycle of stress applications there were discovered statistically reliably lesser amounts of an increase in hydro-cortisone in the blood. Thus, the maximum amounts of the concentration of this hormone, observed at the second and third immobilizations, amounted to (2019±126) n-moles/l and (2142±176) n-moles/l, which is 27.4 and 29.0% ($P<0.05$) less than the corresponding values in monkeys of the first group. Along with this in monkeys of the second group there were noted also lower baseline values of hydro-cortisone, which led to a relative leveling in the amplitude of the increase of this hormone in monkeys of both groups.

Thus, 2 hours after the beginning of the third immobilization the hydro-cortisone content in the second group of animals was equal to (2142±176) n-moles/l, which amounted to 261% of the relative initial values, while in monkeys of the first group the increase in the level of the hormone in response to an analogous application amounted only to 245%, in spite of the significantly higher ($P<0.05$) amounts of its concentration--(2782±139) n-moles/l. The preliminary stress application also contributed to a more rapid restoration of hydro-cortisone in the blood, which was manifest in the reduction of its concentration to the initial level only 6 hours after the beginning of the first immobilization of the stress cycle, while in monkeys of the first group a complete restoration of the hormone in the blood in the indicated period of time occurred only after the fourth application.

Under an increase in the duration of the intervals between immobilizations up to 72 hours, there was observed a progressing reduction of the peak amounts of hydrocortisone concentration, the content of which by the fourth immobilization was reliably no different from the initial level, registered before the beginning of the application (Illustration 9c). In monkeys of this group there was also noted a more pronounced reduction in the baseline level of hydrocortisone, the amount of which before the beginning of the third immobilization amounted to (521±84) n-moles/l, which was less than ($P<0.001$) the initial values by 1.9 times. A complete restoration of hydrocortisone content in the third group of monkeys was observed 6 hours after the beginning of the second immobilization, while after the cessation of the stress applications its content remained somewhat reduced for a period of three days.

The dynamics of the change of the concentration of 11-desoxyhydrocortisone in the blood throughout the entire period of stress applications in monkeys in the first series of experiments on the whole repeated the nature of the response reactions established for hydrocortisone, however it differed by a larger amplitude of the increase (Illustration 10,a). Thus, the maximum growth of the concentration of 11-desoxyhydrocortisone, noted in the 2-3 immobilizations, amounted to 304 and 270%, which was greater than the initial level ($P<0.01$) by approximately 4 times. During an analysis of the curves of the content of 11-desoxyhydrocortisone in the blood in the two other experimental groups of monkeys there were revealed significant differences in the dynamics of this predecessor by comparison with its response reaction noted among animals of the first group (Illustration 10,a,b). In monkeys subjected to a preliminary 2-hour immobilization, throughout the entire cycle of repeated immobilizations, there was discovered a sharp decrease in the amplitudes of the increase of 11-desoxyhydrocortisone, the amounts of which turned out to be 2 times less than the corresponding peak values registered in the first experiment ($P<0.001$) (Illustration 10,b). In the third series of experiments, with the increase in intervals between stress applications to 72 hours, there was noted a pronounced and rapid decrease in the peak values of 11-desoxyhydrocortisone with an absence of statistically reliable differences in the amounts of its increase relative to the initial values, beginning with the fourth immobilization (Illustration 10,c).

Taking into account that all our experiments used one and the same type of stress application-immobilization--produced under strictly standard conditions on clinically healthy, intact monkeys having similar age and mass characteristics, and also the fact that in all experimental groups the hormonal reaction of the adrenal cortex to the first immobilization was comparable both in the absolute amounts, and in the nature of the response dynamic of the steroids in the blood, it seems obvious that the pronounced differences in the functional state of the GGAKS manifested in the three groups of monkeys, are caused by a change in the length of intervals between individual stress applications.

In confirming the relationship we have established there is a certain significance in the detailed study of the change of the content of 17-hydroxylized predecessors of hydrocortisone, even more so since the nature of their dynamic under stress, especially repeated stress, at the present time remains to be studied. It is well known that 17-hydroxylized steroids--derivatives of progesterone and pregnenolone, are common predecessors in the system of the biosynthesis of steroid hormones both in the adrenal and in the testes. It is felt that in male individuals in the formation of the content of 17-oxypregnenolone in the peripheral blood the main role belongs to the adrenals, while of 17-oxyprogesterone--to the testes (Yudaev, et al., 1977; Goncharov, et al., 1977a, 1978b). In addition there is proof that the ranges of the level of 17-oxyprogesterone in the blood are close in phase to the rhythm of hydrocortisone (Gutai, et al., 1977). Analogous data have been obtained in the research of Tavadyan (1981), which established a high correlation of the level of 17-hydroxylized predecessors with the daily and seasonal dynamics of hydrocortisone in rhesus macaques.

While studying the C-rhythms of the steroid hormones in Papio hamadryus there was also discovered a high correlation in the dynamics of 11-desoxyhydrocortisone, 17-oxypregnenolone and 17-oxyprogesterone with hydrocortisone (Taranov, 1981). In addition the absence of a correlation in the daily rhythms of the dynamics of hydrocortisone with progesterone and other predecessors (Taranov, Goncharov, 1981) along with data obtained while studying the content of a wide spectrum of steroids in the blood, flowing directly from the adrenals and testes (Goncharov, et al., 1978b), and along with the nature of the change in the dynamics of steroids during a severe stress allowed the authors to assume that the main path of biosynthesis, leading to the formation of hydrocortisone in papio hamadryus, just as in man, is delta5 way--a synthesis of hydrocortisone through 17-oxypregnenolone.

The simultaneous determination of the content of hydrocortisone and its predecessors in the blood allows the evaluation of the nature of the partial adaptation of various stages of cortico-steroid-genesis and the revelation of its weaker places under conditions of chronic stress. Such an approach gives the possibility to also approach an understanding of the autoregulatory mechanisms of the adreno-cortical activity under stress and is promising in the development of criteria to evaluate the reserve abilities and make a prognosis for the functional activity of the pituitary-adreno-cortical system under conditions of the frequent influence of stress irritants. This is particularly important, since the restructuring and perfection of the mechanisms of corticosteroid-genesis, especially under conditions of the repeated effect of stress, has an exceptionally complex nature and depends, as was previously shown, on many factors, including the length of interval between repeated stress applications.

As was established in our experiments, the dynamics of the content of 17-hydroxylized predecessors of hydrocortisones during numerous immobilizations had distinguishing characteristics in comparison with hydrocortisone and with 11-desoxyhydrocortisone, which were manifest in the large values of the increase of their concentration at each stress application and in the more rapid restoration of these predecessors after the end of the influence of the stress irritant (Illustration 11). In response to the first immobilization there was observed a maximum increase in the level of 17-oxyprogesterone in the blood, an increase in the concentration of which 2 hours after the beginning of immobilization amounted to 500% (Illustration 11,a). As opposed to hydrocortisone and 11-desoxyhydrocortisone the content of 17-oxyprogesterone 6 hours after the beginning of the stress application in monkeys of the first group was significantly reduced, however, it was still reliably higher than ($P<0.001$) the initial values. At the second and third immobilizations there was noted a certain reduction in the maximal amounts of the content of this predecessor, while beginning with the fourth immobilization there was observed a more pronounced reduction of its peak values. The restoration of the level of 17-oxyprogresterone after the end of the stress application occurred more quickly in comparison with hydrocortisone and 11-desoxyhydrocortisone, and within 6 hours after the beginning of the first immobilization there was observed a significant reduction in its concentration, while beginning with the third application and later--the content of 17-oxyprogesterone at 6-hour intervals corresponded to the initial level.

The dynamics of the content of 17-oxypregnenolone in the blood was distinguished by a particularly sharp increase in its level at all stress applications with a maximum at the third immobilization, amounting to a 10-fold increase in the concentration of this predecessor relative to the initial level (Illustration 11,a). After reaching the maximum response there was noted a certain reduction in the peak amounts of 17-oxypregnenolone, which was more pronounced at the final immobilization. However, the amount of the increase of its concentration remained very significant and amounted to 573%.

Differences in the dynamics of 17-oxyprogesterone during preliminary immobilizations of monkeys by comparison with the first group were manifest in a significant reduction of the amounts of the increase of this predecessor and in its more rapid restoration to the initial amounts (Illustration 11,b). Thus, 2 hours after the first immobilization of the stress cycle the concentration of 17-oxyprogresterone amounted to (2.66±0.21) n-moles/l, while after 6 hours-- (0.49±0.09) n-moles/l and was not reliably different from the initial level.

The dynamics of the content of 17-oxypregnenolone in the blood throughout the entire stress cycle, conducted after preliminary immobilization, had a great similarity in the nature of its response reaction with animals of the first group (Illustration 11,b). However, the restoration of the content of this predecessor to the initial values during preliminary immobilization of monkeys took place after the end of the first immobilization of the stress cycle, while in the first group the complete restoration was observed only after the third stress application.

Under conditions of repeated emotional stress with intervals of 2 days between immobilizations there was noted a more pronounced nature of a drop in the amplitude of the increase of 17-oxyprogesterone in comparison with the dynamics of this predecessor in monkeys of the first and second groups, and at the fifth stress application its peak values were not different from the initial level (Illustration 11,c). Along with a sharp drop in the peak values of the concentration of 17-oxyprogesterone there was noted a decrease of its baseline value, however the degree of expression of this process was not different from analogous changes of the baseline dynamics of this predecessor in monkeys of the second group.

The increase of the length of intervals between immobilizations had a significant influence on the dynamics of the content of 17-oxypregnenolone as well, the level of which at the fifth application was 1.9 times less in comparison with its reaction at the first immobilization (Illustration 11,c). Such a significant reduction in peak amounts of this predecessor was not noted in monkeys of the first two groups. A complete restoration of the concentration of 17-oxypregnenolone and 17-oxyprogesterone in the blood to the initial values occurred simultaneously with the normalization of the level in the blood of hydrocortisone and 11-desoxyhydrocortisone, and was registered 6 hours after the beginning of the second and all subsequent immobilizations.

The research conducted allowed us to establish that the more pronounced increase of the adreno-cortical activity occurs during a daily influence of stress irritants, and in comparison with severe stress, leads to a more significant and sharp increase in the level of gluco-corticoids in the blood. Conducting preliminary immobilizations 3 days before an analogous cycle of stress application reduces the peak amounts of the

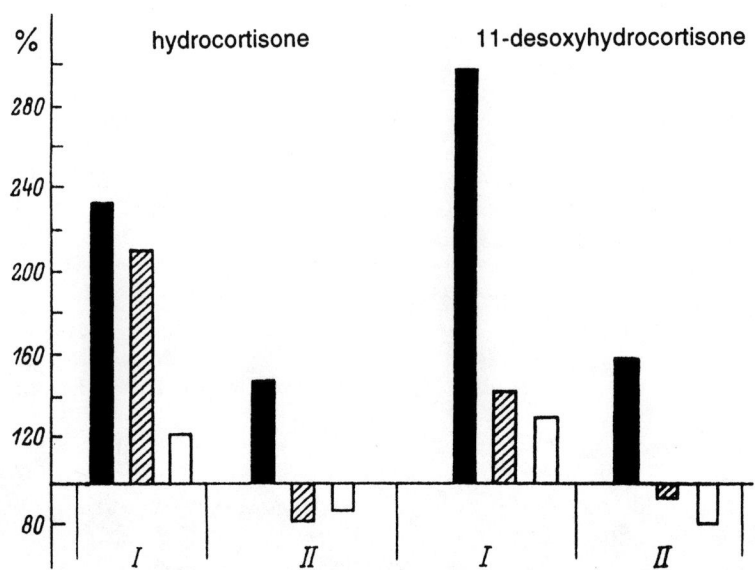

Illustration 12. Degree of expression of the adreno-cortical reaction in monkeys, depending on the regiment of stress stimulation application.

Average amounts of hormone content after 2 hours (I), and 6 hours (II) from the start of all stress applications, % of the initial level; solid columns - cycle of daily 2-hour immobilizations for a period of 6 days; striped columns - analogous cycle of stress applications with preliminary immobilizations; clear columns - series of 2-hour immobilizations, with 2, 24-hour intervals between individual applications.

concentration of gluco-corticoids at the repeated stress applications and aids the increase in the processes of normalization of their levels in the blood, while lengthening the intervals between repeat immobilizations up to 72 hours leads to a rapid suppression of the adreno-cortical reaction.

The established relationship of the nature of the response reaction of the GGAKS to a schedule of application of stress irritants is obviously manifest both in the change of the degree of the secretory activity of the adrenal cortex, analyzed according to peak values of gluco-corticoids and in the dynamics of the change of the process of slowing down its activity, studied according to the restoration of the level of hydrocortisone and its predecessors in periods following repeated immobilizations. (Illustration 12). During an evaluation of the functional state of the GGAKS in monkeys under conditions of the repeated action of stress irritants with various regimens of their application we considered data dealing with the change in the amounts of the ratios of 11-desoxyhydrocortisone/hydrocortisone, 17-oxyprogesterone/11-desoxyhydrocortisone and 17-pregnenolone/11-desoxyhydrocortisone, which are relative indicators of the activity of the enzyme systems of steroid-genesis--11-beta-hydroxylases and 21-hydroxylase of the 17-oxyprogesterone and 17-oxypregnenolone.

An analysis of the amounts of these ratios along with the data offered on the dynamics of the content of hydrocortisone and its predecessors allows us to assume that in response to the first immobilization, evidently, there took place a predominant increase of the initial stages of steroid-genesis in the adrenals: the separation of the side chain of cholesterol, 17-hydroxylation of pregnenolone and progesterone--in comparison with later stages of the transformation of steroids. This is also confirmed by a similar dynamic of the accumulation of 17-hydroxylized predecessors in the blood and by the reduction of the activity of the corresponding enzymes, observed in the initial period of severe stress in another species of monkey--rhesus macaques (Tavadyan, 1981). As opposed to severe stress under repeated ES there occurred an increase in the biosynthesis of steroids at the later stages, in particular, in the phase of the transformations: 17-oxypregnenolone to 17-oxyprogesterone to 11-desoxyhydrocortisone to hydrocortisone.

The revealed changes in the activity of enzymes on the basis of the analysis of the amounts of the ratios of hydrocortisones and its predecessors, along with the dynamic of the content in the blood of the steroids researched, reflects a compensatory restructuring of steroid-genesis under conditions of chronic stress, aimed at the more complete utilization of predecessors of hydrocortisone and can be explained by the activity of repeated bursts of endogenous AKTG. This is confirmed by experiments with the repeated introduction of exogenous adreno-cortical-tropic hormone, leading to an increase in the activity of enzymes of the later stages of cortico-steroid-genesis and to the restructuring of the activity of the gland to a regimen of a more complete utilization of predecessors for the synthesis of hydrocortisone (Fuchs-Hammoser, et al., 1978; McKenna, et al., 1979).

At the same time a comparative analysis of the dynamic of the content of hydrocortisone and its predecessors in the blood, with consideration of the change in the amount of their ratio, indicating the adaptive nature of the restructuring on hormone synthesis during repeated action of stress irritants, attests to the fact that under conditions of repeated ES, produced in a setting of preliminary stress application, just as under the increase of the length of intervals between immobilizations, the increase in reserve abilities of the GGAKS occurred during a less intense cortico-steroid-genesis, than what took place in the first series of experiments. The greater expression of adaptive phenomena of the functioning of the GGAKS in monkeys in the second and third series of the experiments is shown also by data of a correlational analysis.

In all groups of animals, beginning with the first application and later, there was observed a high positive correlation ($R=+0.95+0.99$ with less than 0.05) between hydrocortisone and 11-desoxyhydrocortisone, and also between 17-hydroxylized predecessors. However, in response to the first immobilization in all monkeys a correlative relation between hydrocortisone and 17-hydroxylized predecessors was absent. A statistically reliable positive correlation between them was noted in the first group of monkeys only at the fourth stress application. An analogous situation took place in this group during the analysis of correlational interrelations between 11-desoxyhydrocortisone and 17-hydroxylized predecessors. In contrast among the monkeys of the second group beginning even with the first immobilization with the stress cycle there were established high positive correlative relations between hydrocortisone and 17-oxypregnenolone and 17-oxyprogesterone ($R=0.99$ with $P<0.05$), and also between the latter and 11-desoxyhydrocortisone. A similar dynamic of correlative relations between hydrocortisone and 11-desoxyhydrocortisone and 17-hydroxylized predecessors was observed among monkeys of the third group.

Thus, in our experiments under the influence of repeated stress irritants there was noted a pronounced restructuring of the functioning of the GGAKS, which was manifested in the increase of its reserve abilities and in the change of the amount of the response reaction, as well as in the shortening of the time period necessary to reach the maximum response and in the development of a more rapid slow-down of the adreno-cortical activity. Our data agree with the results of data obtained in other laboratory animals, in which it shows that numerous repeated stress irritants lead to a reduction in the increase of the level of cortico-steroids in the blood and to a significant reduction in the time of their response reaction and restoration (Shorin, Obut, 1973; Shorin, et al., 1975; Shorin, 1977). Such a dynamic of the change of the adrenocortical activity is evaluated as a decrease in the reactivity and an increase in the lability of the GGAKS under these conditions. However, as shown in our experiments, a change in the reactivity of the system and an increase in its lability depend also on the regimen of the application of stress irritants.

The maximum reactivity of the GGAKS to repeated immobilizations we noted during the daily application of the stress. What's more a high level of hydrocortisone and its predecessors in the blood considering the data of the dynamic and the ratio of the steroids allows us to assume that the adaptation of the hypothalamus-pituitary-adreno-cortical system to the activity of the stress was reached through extreme tension of the functional activity of this system. This fact accords with Viru's concept (1981) of the need for a near-maximum tension of the function of the corticocytes for the development of the adaptation of GGAKS to the action of the stress.

However, if you consider that a more adequate criteria of the adaptation to the stress influence is a reduction of its reactivity (a reduction in the amount of the increase of the level of hydrocortisone and its predecessors) and an increase in lability (a hastening of the response and a decrease in the length of the period after the effect), then in the monkeys of the second and third group the process of adaptation was more pronounced. This was manifested in significantly lower amounts of the increase of gluco-corticoids, in a more rapid and complete restoration of their level and in the early development of a high correlation between hydrocortisone and its predecessors in response to repeated immobilizations. Furthermore the increase in the length of intervals between repeated applications led, in monkeys of the third group, to a more pronounced suppression of adreno-cortical activity. Consequently, along with such factors as the strength and length of the stress action, determining the degree of the activation of GGAKS, great significance in the nature of its response under conditions of the repeated action of a single-strength stress is given to the time factor, i.e., the length of the periods between application of the irritant. The increase in the length of these periods acts contrary to the activating influence of the stress, aiding the reduction in the amount of the response reaction of the GGAKS.

The application of preliminary immobilization with a 3-day period after the action, and the increase in the length of intervals up to 72 hours between stress applications create, evidently, conditions for a more optimum interrelation of the central (KRF and AKTG) and peripheral (gluco-corticoid) complexes of the GGAKS, whose functional state, in its turn, is influenced by other regulatory neuro-endocrine systems (Jones, et al., 1984) under conditions of a longer period after the action there is created an opportunity for complete restoration of the functional state of the central adrenergic and other neuro-mediator systems of the brain, which take part in the activation and braking of the GGAKS. In addition, considering the important role in the regulation of the level of gluco-corticoids in the blood in the processes of the interaction of steroids with protein plasma, their bonding with receptors of effector organs, the utilization and metabolic transformations of these hormones in the periphery (Sapronov, 1980; Viru, 1981; Robu, 1982), one can assume the difference of the influence of these factors on the development and degree of expression of the adaptive restructuring of the response reactions of the GGAKS during repeated ES in the 3 groups of monkeys depending on the regimen of the application of stress stimuli.

Chapter 4

THE NATURE OF THE CHANGE OF THE ENDOCRINE FUNCTION OF THE TESTES IN MONKEYS UNDER CONDITIONS OF EMOTIONAL STRESS

Lately the attention of many researchers is drawn to the questions of endocrine regulation of the reproductive function in man and animals under conditions of stress. Of special interest for the resolution of these tasks are data of studies on the clarification of the direction and the dynamic of the change of the content of sex steroids in the blood under the influence of various stress stimulai.

Available data of the literature indicate that suppression of the endocrine function of male sex glands is noted under the widest variety of changes of the external and internal environment of the organism and is one of the more typical manifestations of the stress reaction. Thus, a significant decrease in the level of testosterone in the blood in experimental animals, including monkeys, was revealed under the influence of severe and chronic psychogenic factors (Rose, et al., 1972; Bernstein, et al., 1978), under hypokinesia (Tavadyan, Goncharov, 1981), starvation (Pirke, Spura, 1980), hypothermia (Collu, et al., 1979), singular and repeated immobilizations (Repcekova, Mikulay, 1977; Aleshin, Bondarenko, 1982; Katsia, et al., 1984b), traumatic applications (Gray, et al., 1978), etc.. It has been established that suppression of the endocrine function of the testes in papio hamadryus under conditions of severe emotional stress according to the degree of intensity and duration of the response exceeds the increase on the level of hydrocortisone and its predecessors in the blood (Goncharov, et al., 1978a).

A drop in the concentration of testosterone in the blood among men was revealed during psychosomatic illnesses and myocardial infarction (Gerasimova, et al., 1978; Wang, et al., 1978a; Gubachev, Stabrovsky, 1981), burns, traumas and surgical invasions (Matsumoto, et al., 1970; Aono, et al., 1976; Shurygin, 1980). What's more it was shown that the duration of the suppression of the endocrine function of the testes depends not only on the seriousness of the pathological processes, but also on the size of the operative invasions. During small operations in men suppression of the secretion of testosterone was observed throughout 5-6 days, while with large-scale surgical invasions the period of the reduction in the testosterone concentration in the blood increased up to 3 weeks and coincided with the development of a negative nitrogen balance, reflecting also a catabolic phase of stress (Matsumoto, et al, 1970; Carstensen, et al., 1972; Monden, et al., 1972; Aono, et al., 1976).

Some authors express the opinion that regressive changes in the testes and a reduction in the concentration of testosterone are the same type of adaptive reactions to stress, as the burst of catecholamines and the increase in the level of gluco-corticoids in the blood (Furdui, et al., 1973; Rugal, 1977; Marin, Guragata, 1978).

Along with the theory of the adaptive significance of the suppression of hormonal activity of the testes under stress there is information which indicates that a lengthy reduction in the level of testosterone in the blood can be the reason for a disturbance of the reproductive function, and can also be a factor causing the development of various illnesses, even if the suppression of the androgyn secretion was initially mediated by the illness itself (Garkavi, et al., 1977; Gerasimova, et al., 1978). Experimental studies have shown that gonadectomy leads to a disturbance in the function of the higher nervous activity and vasomotor regulation with the development of post-castration arterial hypertension, and can also lead to a change in the metabolic processes, a disadaptation and extreme increase in the adreno-cortical activity under stress (Vartapetov, Gladkova, 1971; Obut, et al., 1980; Fokin, 1981).

As established in experiments with the cannulation of the testical vein, a drop in the level of testosterone in the blood under stress is caused by the suppression of the steroid-genesis in the testes, and not by an increase in the metabolic clearance of the hormone (Bardin, Peterson, 1967; Aleshin, Bondarenko, 1982). The suppression of androgenopoesis observed under the influence of stress irritants is accompanied by a decrease in the mass of the prostate gland and gonads with the development of regressive morphofunctional changes in the testes all the way to deep structural disturbances of the hormone producing elements and their complete atrophy (Blinova, 1977; Rugal, 1977; Repcekova, Mikulay, 1977; Aleshin, Bondarenko, 1982).

At the present time it has been established that for the study of the endocrine function of the reproductive system under the influence of various stress factors and during pathological states of the organism the most appropriate research objects are monkeys in view of their great similarity with man in the direction of the testicular steroid-genesis and the spectrum of the secreted hormones, in the mechanisms of the regulation of the hypothalamus-pituitary-gonad system (GGGS) and the presence of close functional interrelationships between the adrenal glands and the testes (Goncharov, et al., 1977b; Yudaev, et al., 1977; Yamamoto, et al., 1978; Tavadyan, 1981; Tavadyan, Goncharov, 1981; Taranov, 1981; Katsia, et al., 1984b).

Characteristic for sexually mature Papio hamadryus males are single-phase C-rhythms of the secretion of testosterone with an acrophase during the evening and with minimal content of the hormone in the morning hours, and furthermore a maximum concentration in the blood of this species of monkey is noted in the summer-fall months with a subsequent gradual reduction during the winter (Taranov, 1981). It has been established that various situational factors lead in monkeys to the disturbance of normal daily rhythms of testosterone secretion, the restoration of which occurs according to the adaptation of the animals to the conditions of the experiment (Plant, et al, 1974; Rose, et al., 1978; Steiner, et al., 1980; Tavadyan, 1981). These studies show a clear interrelation between the C-rhythm of the testosterone content in the blood and change in the level of the luteinizing hormone (LG), the increased frequency of the secretion of which precedes a greater increase in the concentration of testosterone.

In its turn, the gonadotropic function of the pituitary is under the control of the hypothalamus releasing factor for the luteinizing hormone--luliberin (LG-RG), and furthermore the balance of the interrelationship between the hypothalamus-pituitary and the testes is realized by means of the mechanisms of the long (androgyn) and short (gonadotropin) reaction (Yudaev, et al., 1976; Veropotvelyan, et al., 1980; Babichev, 1981; Le Tkhu Lien, Chernositov, 1983). An important factor determining the nature of the reaction of GGGS under stress is the initial level of the secretory activity of the testes (Davies, et al., 1979). Thus, a lengthy fixation of male rhesus macaques, produced in a setting of maximum seasonal activity of the sex glands, led to a drop in the level of testosterone in the blood, while under conditions of analogous stress application, coinciding with a period of low secretion of androgen's, there was noted an increase (by 150%) of the concentration of the hormone in the blood (Tavadyan, 1981; Tavadyan, Goncharov, 1981). Similar data were obtained in experiments using the immobilization of male papio hamadryus, in whom there was revealed an absence of the inhibiting effect of stress on the hormonal activity of the testes with a low initial content of testosterone in the blood (Taranov, 1981; Catsia, et al., 1984b). It should be noted that the indicated differences in the change of the endocrine function of the gonads were observed only at the initial stages of the action of the stress irritants, while the further direction of the response reaction of the GGGS under stress, regardless of the

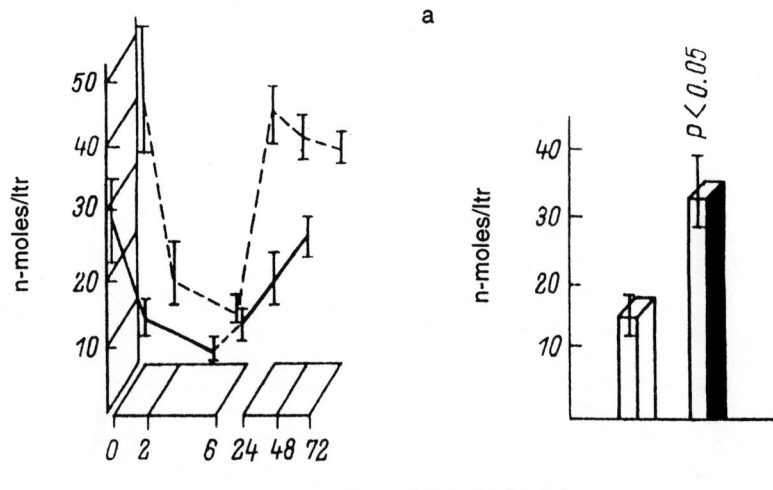
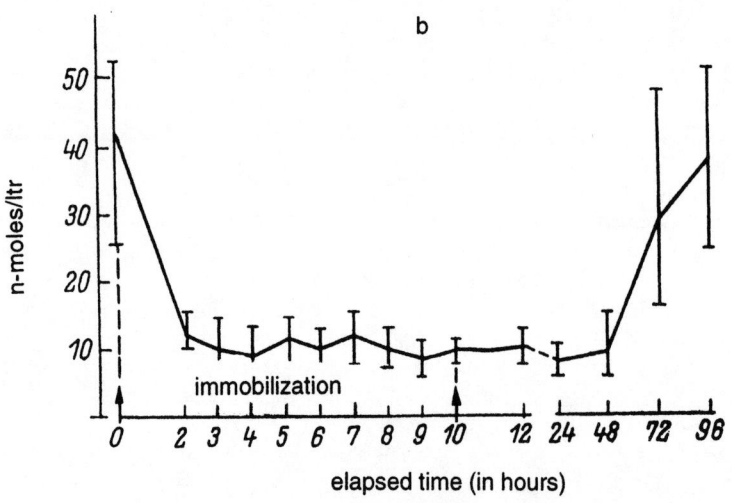

Illustration 13. Dynamics of testosterone content in the blood during severe (a) and prolonged (b) emotional stress. Mathematical averages with reliable intervals.

Y-axis - testosterone content, n-moles/ltr; X-axis - time of blood draw: O - before immobilization; 2,6,24,48,72 - corresponding periods of time from the start of immobilization; solid line - stress in the morning (10:00-12:00, dotted line - stress in the evening (22:00-24:00). Amplitude of testosterone drop under stress in the morning (blank columns) and evening (solid columns).

initial secretory activity of the testes, was characterized by a reduction in the concentration of testosterone in the blood.

In experiments we conducted on monkeys to study the dynamics of the content of testosterone in the blood under ES it was shown that the duration and degree of the expression of the suppression of the endocrine function of the testes was clearly determined by the force and duration of the action of the stress factors, by the phase of the daily rhythm of androgen secretion at the moment of application, and also by the duration of the intervals between the application of repeated irritants. As shown in Illustration 13,a, the maximum reduction in testosterone concentration among sexually mature papio hamadryus males during severe ES, produced during the morning hours, occurred 6 hours after the beginning of the stress application. The content of the hormone in the blood at that moment amounted to (8.7±1.8) n-moles/l, which was less than the initial level by 71% ($P<0.0001$). During the period of the after effect of stress the testosterone content in the blood gradually was restored, reaching the initial values on the second and third day.

Similar data were obtained earlier by Goncharov, et al. (1977a) and Taranov (1981), who established that along with a drop in the testosterone concentration under conditions of severe stress there occurs in monkeys a significant reduction in the level of its main metabolite-- 5-alpha-dihydrotestosterone, the dynamics of which have a high correlation with a change in testosterone content.

Immobilization during the evening, during the phase of high secretory activity of the testes, lead to a greater drop in the concentration of testosterone in the blood (in percentage terms) (Illustration 13,a). The size of the amplitude of the reduction of testosterone content in the blood amounted to (29.8+3.9) n-moles/l, which was 1.7 times greater than ($P<0.05$) the amplitude of the drop in the hormone observed during immobilization in the morning hours-- the phase of the low daily secretion of androgens. However the restoration of the secretory activity of the testes in monkeys in the period of the after effect of stress, produced during the evening, occurred more rapidly and was noted within 4 hours of the beginning of the stress application (Illustration 13,a).

Thus, a more significant drop in the concentration of testosterone in the blood during ES in the evening hours (in percentages) occurs in a setting of a high baseline level of the hormone. Consequently, as opposed to GGAKS to strengthen the reactivity of GGGS, which has an opposite reaction toward suppressing the endocrine function of the testes, a high initial level of androgen production is necessary. The single direction of the change of hormonal activity of the testes toward suppression of androgenopoesis during immobilization of monkeys at various times of the day is caused, in our opinion, by sufficiently high level of androgren secretion in these animals both during the evening and the morning hours, which in both cases was also the basis for the manifestation of the characteristic reaction of the testes to the action of the stress irritants.

Under the influence of a 10-hour immobilization there was noted in monkeys a deep and prolonged suppression of the hormonal activity of the testes. Throughout the entire period of immobilization there were registered borderline low values of testosterone content in the blood, amounting to around 25% of the initial amounts. In the first and second days of the period of the aftereffect of ES the concentration of testosterone remained low and amounted to only 21% ($P<0.01$) and 23% ($P<0.01$) of the initial level. The restoration of the hormone content in the blood occurred only on the fourth day of the period of the aftereffect of stress (Illustration 13,b). In spite of the large number of works dedicated to the study of the functional state of GGGS under stress and to the questions of the regulation of this system, concrete mechanisms inhibiting the synthesis and secretion of androgens in the testes during stress nevertheless remain unexplained to the present day. It is assumed that the suppression of the testicular steroid-genesis under stress is caused by a disturbance in the synthesis and secretion of LG in the pituitary and by a slowdown in the release of luliberin in the hypothalamus. This point of view is based on data which indicate that under conditions of severe and chronic stress along with a drop in the concentration of testosterone, there is observed a reduction in the level of LG in the blood and an increase of LG-RG in the hypothalamus, indicating the inhibition of its release (Aono, et al., 1972b; Collu, et al., 1979; Pirke, Spura, 1980; Taranov, 1981).

At the same time a number of works cite date on the increase of LG concentration in the blood under stress, and furthermore the increase in the level of gonadatropine the authors relate to engaging the reaction mechanism as a result of a low content of testosterone in the blood (Monden, et al., 1972; Aono, et al., 1976). The absence of a disturbance in the synthesis and secretion of LG under conditions of stress is also indicated by experiments in which there was discovered a complete preservation of the response of LG to stimulation by exogenic luliberin (Campbella, et al., 1977; Collu, et al., 1979; Tavadyan, 1981; Katsia, et al., 1984). The increase in the level of LG in the blood during a low concentration of testosterone allows us to assume that under the influence of stress irritants there occurs a reduction of the sensitivity of the testes to endogenic LG. To a certain extent this is confirmed by data about the reduction of the testicular response during the introduction of chorionic gonadotropin (chKhG) both during severe and chronic stress (Aono, et al., 1972a; Repcekova, Mikulay, 1977; Tavadyan, 1981).

One of the possible mechanisms of the reduction of the sensitivity of the testes to endogenic LG during stress can be a disturbance of the blood flow in these glands under the influence of excessive release of KA. It has been established that the introduction of catecholamines causes a disturbance of the testicular blood flow, leads to a reduction in the level of testosterone in the blood and to the suppression of its biosynthesis (Levin, et al., 1967; Damber, Johnson, 1978; Verhoven, et al., 1979; Gotz, et al., 1983). Furthermore the introduction of substances which inhibit the spasm of the blood vessels, forestalls the inhibiting effect of stress on the secretion of testosterone (Collu, et al., 1982; Chirkov, 1984).

One point of view exists, according to which a reduction in the level of testosterone in the blood under stress is the result of the inhibiting influence of AKTG and hydrocortisone on the hormonal activity of the testes. This theory is based on data concerning the reduction in the level of testosterone in the blood after the introduction of AKTG and gluco-corticoids (Beitnis, et al., 1973; Irvine, et al., 1974; Forest, 1978; Kim, et al., 1978). The research of Cumming, et al. (1983) shows that an increase in the level of hydrocortisone in the blood in healthy males after the introduction of insulin or hydrocortisone was accompanied by a rapid reduction in the concentration of testosterone, which, in the opinion of the authors, was caused by the direct influence of an elevated concentration of hydrocortisone on the production of androgens in the sex glands.

However, in the literature there is proof of the preservation of the normal secretory activity of the testes in rats after an adrenalectomy, causing an increase in the level of AKTG in the blood, and there's even data on the increase of the concentration of testosterone after the introduction of AKTG and gluco-corticoids (Forest, 1978; Shaison, et al., 1978; Fenske, et al., 1979; Gray, et al., 1978; Lang, et al., 1979). The difference revealed in the time periods for the restoration of the adreno-cortical activity and the secretory activity of the gonads during the post-stress period, when the content of AKTG and gluco-corticoids reached its initial values, while the level of testosterone remains reduced, may also attest to the absence of the inhibiting influence of the activation of GGAKS on the endocrine function of the testes (Monden, et al., 1972; Repcekova, Mikulay, 1977; Gray, et al., 1978; Shurygin, 1980; Chirkov, 1984).

The research of Tsulay (1985) shows that the introduction of metopyrone 30 minutes before the beginning of a 2-hour immobilization prevented, in sexually mature papio hamadryus males, the increase of hydrocortisone concentration in the blood, and even reduced its level, however, it did not remove the inhibiting influence of stress on the endocrine function of the testes, which, in the opinion of the author, indicates the absence of a cause and effect relationship between the increase in the level of hydrocortisone and a drop in the concentration of testosterone in the blood during severe stress.

It is felt that in the suppression of androgen production under stress and, in particular, in the reduction of the sensitivity of the testes to LG, a certain role may belong to the strengthening of the development of the pituitary prolactin (Ajika, et al., 1972; Hafiez, et al., 1972; Euker, et al., 1975). In addition, the prolonged influence of the stress stimulants (cold, immobilization, physical loads) in mice revealed a reduction in the concentration of prolactin in the blood (Du Ruisseau, et al., 1978).

Thus, the neuroendocrine mechanisms of the suppression of the endocrine function of the testes under conditions of stress remain so far unexplained, while data on the study of the secretion of androgens under conditions of chronic stress is limited and of a contradictory nature. Thus, during lengthy and repeated immobilization in rats there occurs a deeper reduction in the level of testosterone in the blood as compared to the less prolonged single applications (Repcekova, Mikulay, 1977). However, in male rhesus macaques under conditions of chronic stress (the flight reactions) there was observed a rapid restoration of the secretory activity of the testes, and furthermore the level of testosterone at the moment of the cessation of repeated stress applications was even somewhat higher than the initial value (Mason, et al., 1968). As our studies showed, under conditions of the daily influence of stress irritants in sexually mature papio hamadryus males (first series of experiments) there occurs a significant and prolonged suppression of the endocrine function of the testes (Illustration 14, a).

Thus, after an incomplete restoration of the level of testosterone in the blood, observed 24 hours after the beginning of the first immobilization in response to the second stress application, there occurred an even greater reduction in the level of androgen in the blood, which throughout the subsequent immobilizations remained at a borderline low level. It should be noted that in a setting of low concentrations of testosterone the dynamics of its response changed, beginning with the third immobilization: 2 hours after the beginning of the stress application there was observed a brief and insignificant increase in the level of androgens in the blood, which turned out to be statistically reliable ($P<0.01$) for the third and fourth immobilizations.

Thus, the results of this series of experiments showed that a daily 2-hour immobilization of monkeys throughout a period of 6 days leads to a deep and persistent suppression of the endocrine function of the testes, persisting throughout the entire stress cycle and for 2 days of the post-stress period.

The Nature of the Change of the Endocrine Function... 87

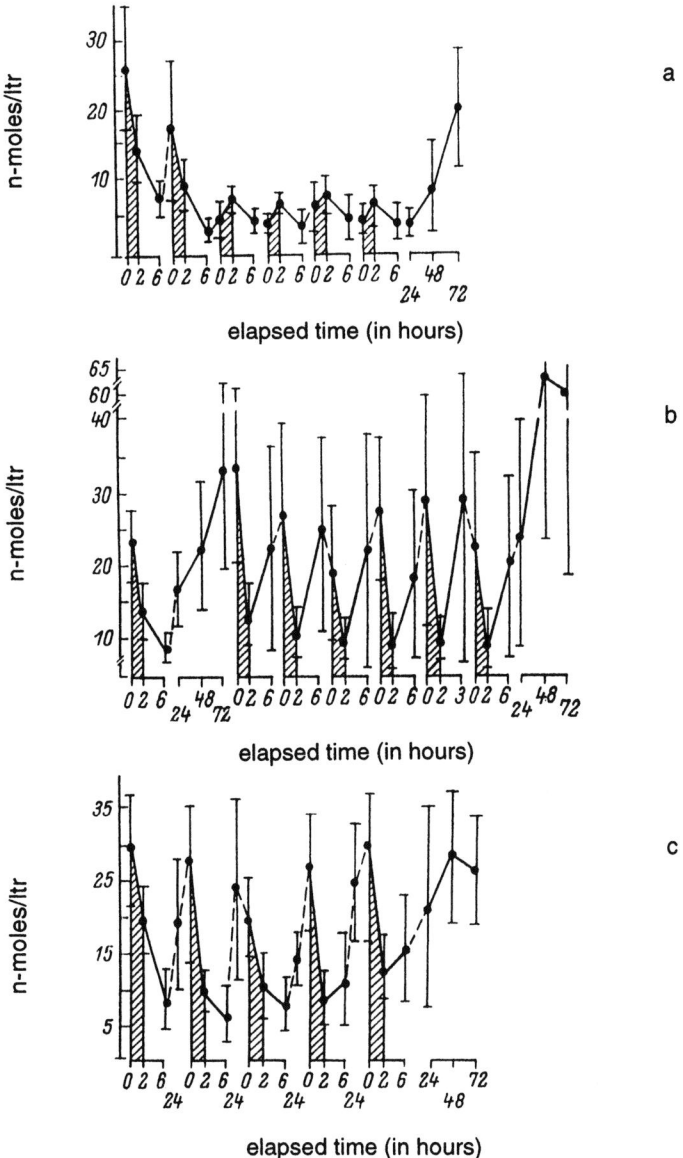

Illustration 14. Nature of the change in testosterone content in the blood of monkeys under repeated emotional stress. Mathematical averages with reliable intervals.

Key: same as Illustration 9.

Illustration 14,b presents the dynamics of the content of testosterone in the blood under conditions of an analogous stress cycle, but with a preceding (3 days earlier) immobilization. The dynamics of the reduction of the level of testosterone in response to preliminary immobilization were similar to the response dynamics to the first immobilization in animals of the first series of experiments. Minimal values of the hormone concentration were observed 6 hours after the beginning of the stress application and amounted on the average to 30% of the initial level. At the second day the concentration of testosterone returned to the initial value. During subsequent immobilizations of the 6 day stress cycle the curve of the dynamics of the testosterone level was characterized by a sharp reduction for 2 hours after the beginning of the immobilization (on the average, by 70%) and by a subsequent rapid and complete restoration of androgen concentration 6 hours after the beginning of each stress application. On the second and third days of the post-stress period in monkeys in the second series of experiments there was noted a sharp increase in testosterone content in the blood up to amounts which exceeded the normal level by 2.6 times.

With the increase in the length of the intervals between stress applications in the third series of experiments in all monkeys there was noted a transitory suppression of the secretory activity of the testes with the greatest decrease in testosterone content 2-6 hours after the beginning of the repeat immobilizations and with a subsequent restoration of its level in 24 hours (Illustration 14,c).

As a result of the experiments conducted there were revealed 2 types of endocrine reaction of the testes to intermittent immobilization stress: a lengthy, many-day suppression and a daily transitory 2-hour drop in the secretory activity. Consequently, the possibility for the development of adaptive restructurings of the function of GGAKS under intermittent emotional stress, and of the sympatho-adrenal and hypothalamus-pituitary-adreno-cortical systems, is related to the regimen of the application of the stress irritants. Preliminary immobilization and the increase of intervals between applications has an adapting nature and promotes a high reactivity of the hypothalamus-pituitary-gonad system during repeated ES.

It should be noted that the suppression of the secretory activity of the testes in monkeys, discovered under conditions of daily immobilizations, according to its depth and duration, is comparable only to the suppression of the testicular secretion of androgens in males during extensive surgical invasions (Aono, et al., 1972a, b; Wang, et al., 1978b). In spite of the fact that the development of regressive morphofunctional changes in the testes and the suppression of their hormonal function is seen as the manifestation of an adaption to stress (Furdui, et al., 1973; Blinova, 1977; Rugal, 1977; Marin, Guragata, 1978), the excessive and lengthy suppression of the secretory activity of these glands under the influence of psychoemotional irritants is a negative factor and can lead to a disturbance/breakdown not only of the reproductive function of the organism, but can serve as a contributing factor in the pathogenesis of neurogenic illnesses. Consequently, the development of adaptation to the influence of stress irritants in the first series of experiments was reached at a high "cost"--the excessive activation of the SAS, the GGAKS and a lengthy suppression of androgens.

However, the increase in the resistance to ES by means of high energetic expenditures and extreme stress of the neuro-endocrine systems is not expedient, since it can lead to a breakdown in the mechanisms of adaptation, the de-training of other functional systems of the organism and a reduction of their structural and energetic resources (Garkavi, et al., 1977; Meerson, 1981). Therefore of special interest are the results of the second and third series of the experiments, which indicate the possibility of an increase in the resistance of the self-regulating mechanisms of the neuroendocrine systems, aided by the increase of the duration of the intervals between emotiogenic situations. The reduction in the degree of the suppression of the endocrine function of the testes under these conditions is an experimental basis for the possibility of the development of rational regimens of the application of stress stimulai, for the purpose of increasing the anabolic fund of the organism in man and animals under lengthy psychoemotional loads.

FUNCTIONAL TESTS WITH LULIBERIN AND CHORIONIC GONADOTROPIN IN MONKEYS UNDER CONDITIONS OF PHYSIOLOGICAL CALM AND STRESS

As we can see from this brief review of the literature, at the present time the question remains unresolved whether or not a reduction in the testosterone concentration in the blood under the influence of stress factors is the result of a suppression of the function of the hypothalamus-pituitary complex or is caused by a disturbance on the level of the steroid producing gland--the testes. The use of functional tests with luliberin (LG-RG) and CKG allows us to differentiate those disturbances arising in the central branches of the system--the hypothalamus and pituitary, from disturbances in the peripheral target glands--the gonads.

The introduction of LG-RG and CKG is a commonly accepted test for the study of the functional reserves of the gonadatropic function of the pituitary and the sex glands (Arslan, et al., 1976; Lau, et al., 1978; Tokar, 1980). It has been established that LG-RG is a decapeptide (Yudaev, et al., 1976), while the chorionic gonadatropin of man is a heterogeneous glycoprotein compound, consisting of at least 3 fractions with various chemical structures (Dimitrov, 1979), and has a greater similarity with the chorionic gonadatropin of monkeys (Chen Hao Chia, Halgen, 1976).

To study the mechanisms of the suppression of the endocrine function of the testes, together with Katsia (Chirkov, Katsia, 1983; Katsia, et al., 1984a) we conducted 6 series of experiments:

1. Interveinous introduction of 100 mkg of LG-RG per animal;
2. Interveinous introduction of CKG in a dose of 4500 ED per animal;
3. Interveinous introduction of LG-RG in the same dose with a subsequent immobilization;
4. Interveinous introduction of CKG in the same dose with a subsequent immobilization;
5. Control experiments with the introduction of a physiological solution under conditions of physiological calm and
6. Introduction of a physiological solution with subsequent 2-hour immobilization.

For the determination of steroids blood was collected from the elbow vein in amounts of 3-4 ml according to the following schedule: before, 30 minutes, 1.5, 2.5, 4.5, 6.5, 24, 48, and 72 hours after the introduction of the preparation. In the experiments with the introduction of LG-RG and CKG and subsequent stress application, the 2-hour immobilization was begun 30 minutes after the introduction of the preparation.

The choice of the doses of luliberin and the chorionic gonadatropin were made on the basis of the data in literature on the use of LG-RG and CKG as functional tests in various stress situations in man and animals, including monkeys (Nakashima, et al., 1975; Campbella, et al., 1977; Tache, et al., 1978, 1980; Toivola, et al., 1978; VanLoon, 1978; Collu, et al., 1979; Tokar, 1980; Taranov, 1981; Taranov, Goncharov, 1981).

With the introduction of LG-RG in intact sexually mature male papio hamadryus the maximum increase in the concentration of testosterone in the blood was noted 1.5 hours after the introduction of the preparation and amounted, on the average, to 80%, but within 2.5 hours its concentration did not reliably differ from the initial level (Illustration 15). The introduction of LG-RG 30 minutes before the beginning of the stress application did not prevent the reduction in the level of testosterone in the blood, caused by stress, however the rate of the reduction of the concentration did slow down and its content within 1.5, 2.5 and 4.5 hours from the moment of the introduction of the preparation was reliably higher ($P<0.01$), than in monkeys in the control experiment with a 2-hour immobilization (Illustration 15). After 6.5 hours there was observed a maximum reduction in the level of testosterone in the blood by 70% in comparison with the initial value.

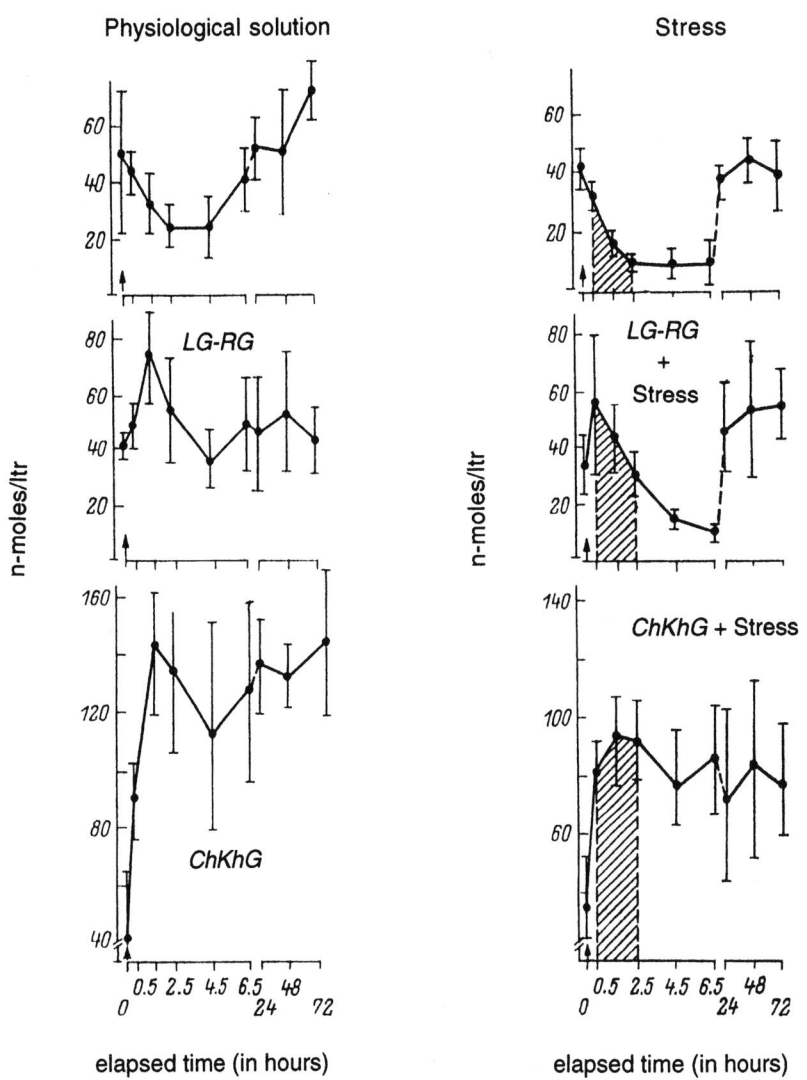

Illustration 15. Effect of luliberin and chorionic gonadotropine on the testosterone level in the blood of monkeys under severe stress. Mathematical averages with reliable intervals.

Y-axis - testosterone concentration, n-moles/ltr; X-axis - time of blood draw: 0 - before the injection: and the elapsed time, in hours, after the injection; arrows - time of injection; LG-RG - releasing factor for a luteinizing hormone (luliberin); ChKhG - chorionic gonadotropine (man): striped sections - period of immobilization.

Table 2. Change in catecholamine excretion in urine among sexually immature male Papio hamadryas during adaptation to confinement in individual cages.

Period of research	Researched compounds (n-moles/24-hrs)		
	A	NA	DA
First week:			
1st 24-hr period	8.34±2.15	32.08±3.68	451.0±77.66
2nd " "	6.99±1.44	30.02±4.16	409.1±89.50
3rd " "	5.64±1.15	27.76±3.09	395.1±87.60
Second week:			
1st 24-hr period	5.68±1.18	21.08±3.04 $P<0.05$	309.6±66.553
3rd " "	6.13±1.44	22.04±2.60 $P<0.05$	328.7±64.22
Third week:			
1st 24-hr period	5.96±1.35	22.88±1.95 $P<0.05$	328.5±64.32
3rd " "	5.07±1.77	20.9±2.89 $P<0.05$	321.8±55.49
Fourth week:			
1st 24-hr period	6.43±1.41	22.55±1.86 $P<0.05$	312.6±42.69
3rd " "	5.21±0.86	21.98±1.67 $P<0.02$	317.0±30.71

Note: P- reliability of differences relative to the first 24-hour period of the first week.

The introduction of CKG to intact monkeys led to a significant increase in the concentration of testosterone which reached (137.1±12.1) n-moles/l at 1.5 hours, which was 200% over the initial values (Illustration 15). A high level of testosterone in the blood under the influence of CKG was noted throughout the following 3 days. The curve of the dynamics of testosterone under conditions of stress with a preliminary introduction of CKG was identical to the curve with the introduction of CKG in intact monkeys (Illustration 15). The maximum increase in the concentration of the hormone in animals subjected to stress in a setting of CKG was noted 1.5 hours after the introduction of the preparation, and its content at this time was 160% greater than the initial level. However, the absolute amounts of the maximum response--(96.5±9.5) n-moles/l--were reliably lower ($P<0.05$), than in intact monkeys with the introduction CKG. After 6.5 hours the concentration of testosterone in the blood amounted to (86.4±9.67) n-moles/l, which exceeded its initial level by 2 times, but during this period and in the following 3 days the content of the hormone was reliably less ($P<0.05$), than in animals with the introduction of CKG.

Thus, conducting the experiments with an introduction of LG-RG and CKG allowed us to establish that under conditions of severe stress in monkeys there occurs a change in the response of the pituitary-testicular system to the stimulating influence of luliberin. In addition, there are data on the preservation or even the increase of the response of the pituitary to the stimulation of LGl-RG both during severe, and during chronic stress (Campbella, et al., 1977; Tache, et al., 1978; Collu, et al., 1979). The contradiction of the data from literature and the results we obtained is evidently related to the means of evaluating the effect of the introduction of luliberin. In the works cited the effect of luliberin was evaluated according to the dynamics of the change concentration of the luteinizing hormone, while in our work--according to the nature of the dynamics of testosterone. In the latter case the resulting change in the concentration of testosterone depended not only on the burst of the endogenic luteinizing hormone, but also on the sensitivity of the gonads to the tropic hormone of the pituitary.

Experiments with a preliminary 2-hour immobilization through the introduction of CKG showed that the preliminary introduction of the tropic hormone not only prevented the suppression of testosterone secretion under stress, but even significantly increased the level of the hormone in the peripheral blood. However, there were also exposed certain differences in the response reaction of the testes to stimulation by a tropic hormone: The increase in the testosterone concentration under the influence of CKG in animals subjected to a 2-hour immobilization, was reliably lower than the response in intact animals. An analogous suppression of the sensitivity of the testes to tropic hormone was revealed in men after surgical invasion (Aono, et al., 1972b).

These results give the basis for the assumption that during stress a relative shortage of gonadotropin is developed, which is compensated for by the introduction of pharmacological doses of CKG. According to the available data, during severe stress the concentration of LG in the blood rises (Aono, et al., 1972b, 1976; Monden, et al., 1972; Euker, et al., 1975). These data support our assumption about the possibility of the development under stress of a relative shortage of LG, which, likely, is related to a reduction in the sensitivity in the testes as a result of a disturbance of their blood flow under the influence of catecholamines (Damber, Janson, 1978). This is confirmed in experiments using substances which prevent blood vessel spasms, the introduction of which prevented the inhibiting effect of the stress applications on the testicle secretion of testosterone (Collu, et al., 1982). A reduction in the sensitivity in the testes to the influence of gonadotropines under conditions of stress (hypokinesia, is shown by the research conducted by Tavadyan (1981) on rhesus macaques. As a contradiction of to these data and the results we obtained in a number of works there was revealed a complete preservation of the response reaction of the testes during the introduction of CKG under stress (Nakashima, et al., 1975; Collu, et al., 1979; Tache, et al., 1980). This difference is related to the method of conducting the experiment. In the works we cited the test with CKG was conducted after the end of the stress application, while in our experiments the influence of CKG on the testes was realized during the immobilization stress.

Thus, based on the data obtained one can assume that during severe stress in male papio hamadryus there occurs a reduction in the sensitivity of the testes to tropic hormones of the pituitary, which, however, does not exclude the possible disturbance of the functional state of the higher regulatory centers--the hypothalamus and the pituitary. The data obtained on the reduction of the gonad sensitivity to tropic hormone of the pituitary allows us to explain the varied nature of the change of the content of LG and testosterone in the blood, revealed during stress (Monden, et al., 1972; Euker, et al., 1975; Aono, et al., 1976), as a reduced sensitivity--under these conditions--of the testes to the stimulating activity of insufficient physiological doses of the endogenous LG.

At the same time the data we obtained on the prevention of a reduction in the level of testosterone in the blood under stress with the help of a preliminary introduction of CKG attests to the possibility of using this preparation for the purposes of preventing disturbances of the endocrine function of the testes in man and animals under extreme conditions.

Chapter 5

THE NEUROENDOCRINE COMPONENTS OF EMOTIONAL STRESS IN SEXUALLY IMMATURE MONKEYS

Least studied at the present time are questions related to the nature and the peculiarities of the formation of the hormonal balance in sexually immature animals under conditions of repeated influence of intense psychogenic irritants. At the same time available data indicate pronounced age differences in the function of the neuroendocrine system, which are manifest not only in various degrees of the secretory activity of the steroid producing gland and in the production of KA under conditions of physiological calm, but also in the nature of the response hormonal reaction to the influence of the stress irritant (Anokhina, 1975; Anisimov, 1979; Zarechny, 1979; Butnev, et al., 1980; Mustafin, Sitdikov, 1980; Obybok, 1981). It has been established that the concentration and distribution of KA in various sections of the myocardium in man depends on age, and furthermore beginning from 3 months there is observed a slow increase in KA content in the heart, the level of which reaches its greatest amounts by the age of 20-30 (Obybok, 1981). During the period of sexual maturation in the majority of animals there occurs an increase of KA in the wall of the blood vessels with the highest level of NA in the kidney arteries (Zarechny, 1979).

The research of Butnev, et al. (1980) has established that the most significant age differences in the nature of the hormonal reaction to stress in sexually immature papio hamadryus males is the less pronounced increase in the level of hydrocortisone and the absence of a drop in the content of androgens in the blood, which in the opinion of the authors, is related to an insufficient functional maturity of the adrenergic structures of the brain, responsible for the hypothalamic control of the pituitary functions, and is also related to a low sensitivity of the testes to gonadotropins. In analyzing available data on the modeling of the neurogenic illnesses in man in experiments on monkeys, it is significant to note that as opposed to adult individuals, in sexually immature animals under the influence of repeated stress irritants there occurs a relatively quick formation of various psychopathological states, including neurogenic arterial hypertension (Startsev, 1971, 1976, 1977; Magakyan, 1977; Dzhalagonia, 1979). Available data indicates that a disturbance in the adrenergic mechanisms of the regulation of the physiological functions is one of the leading factors in the development of psychosomatic pathology, and in particular of neurogenic arterial hypertension (Sudakov, 1976, 1979; Belova, Kvetnansky, 1981; Gubachev, Stabrovsky, 1981; Markel, 1983; Sokolov, Belova, 1983).

It was also established that along with the activation of the adrenergic structures of the brain the development of a persistent arterial hypertension requires "the simultaneous engagement of the hormones of the cortical and medullary layers of the adrenals" (Sudakov, 1976). At the same time in recent years theories about the neurohumorus mechanisms participating in the regulation of the cardiovascular functions have significantly expanded. First and foremost it has been shown that besides catecholamines, gluco-corticoids and mineral-corticoids, an important role in the pathogenesis of AG is given to renin-angiotension-aldosterone system, to the prostaglands, to the substance P, to other vasoactive neuropeptides and to the disturbance of their interrelation. It is also well-known that along with glucocorticoids, which have a permissive effect on the action of KA, great significance in the regulation of the activity of the SAS belongs to the androgens (Bakhova, 1975; Van Loon, 1978; Anisimov, 1979; Chirkov, Goncharov, 1980a; Chirkov, 1984), and furthermore the reduction of the endocrine function of the testes is seen as one of the factors at the basis of the development of the cardiovascular illnesses, including arterial hypertension (Vartapetov, Gladkova, 1971; Gerasimova, et al., 1978; Gubachev, Stabrovsky, 1981; Fokin, 1981).

Considering the important role of hormones and neuromediators in the regulation of the visceral functions and the complex nature of the restructuring of the hormonal balance during chronic stress, as well as the possible significance of the age peculiarities of the neuro-endocrine mechanisms of the development of ES, of significant interest for the clarification of the pathogenesis of psychosomatic disturbances is the research on the detailed study of the dynamics of the response reactions of the SAS and of the steroid-producing glands in sexually immature monkeys during multiple stress applications, even to the appearance of pathological manifestations of stress and their stabilization.

Accordingly, we conducted experiments on 5 sexually immature males, previously adapted to the conditions of confinement in individual cages and to the experimental procedures for a period of 4 weeks. The development of emotional strain was reached with the assistance of a series of 10 cruel 3 hour immobilizations of nonanaesthetized monkeys with intervals of 2 days between the stress applications (Repin, Startsev, 1975; Startsev, Chirkov, 1977). So that the experiment maximally approached experiments on adult animals and also to test the reactions to the threat of capture, before starting all 10 immobilized sexually immature males were subjected to a standard procedure of capture imitation with a 5 minute chase in their living cage, after which they were submitted to the experiment. The study of the neuro-hormonal indicators was carried out during a month in the initial state (becoming accustomed to the setting and the conditions of the experiment), the entire period of the stress applications and a 4 day post period. Arterial pressure (AD) (systolic-AD_s and diastolic-AD) was measured with a cuff according to the Korotkov method.

Measurements of the AD during the adaptation period were made on a daily basis throughout the first week and subsequently with intervals of 1 day. In the experiments with immobilization AD was measured according to this schedule: Before, in 3 and 24 hours after the beginning of each stress application. During the post-period the measurement of AD was conducted on a daily basis throughout the first week and subsequently every 2 days for a period of 6 months. The average pressure (AD_{sr}) was computed according to this formula: $AD_{sr} = AD_d + 0.42 \times AD_{pulse}$), where $AD_{pulse} = AD_s - AD_d$. Registration of the electrocardiogram (EKG) at 12 points of contact was made during the adaptation period, throughout the cycle of the stress applications, before and 3 hours after the beginning of each immobilization, and during the post-period for 6 months (once each month). An evaluation of the T-point dynamics was made according to the indicator of the summary profile of the T-point () at the registered EKG-points of contact (Belkania, 1982).

THE DYNAMICS OF THE CHANGE IN CATECHOLAMINE EXCRETION IN THE URINE OF SEXUALLY IMMATURE MONKEYS UNDER CONDITIONS OF REPEATED EMOTIONAL STRESS

In the process of adapting to the conditions of confinement in individual cages and to the experimental procedures, in sexually immature monkeys, as in adult individuals, there is seen a reduction in the amount of KA excretion, however, in sexually immature animals this process occurs significantly more rapidly and a complete stabilization of the amounts of KA excretion is noted at the end of the first week. In comparison with the amounts of KA excretion registered at the end of the period of adaptation, in sexually immature and adult monkeys there are revealed comparable amounts of NA excretion, while the level of excretion of A ($P<0.05$) and DA in sexually immature animals is somewhat reduced (Illustration 2).

Our studies (Chirkov, Goncharov, 1980a, 1980b; Chirkov, et al., 1986a, 1986b) has established that under conditions of severe ES (the first immobilization) in sexually immature monkeys there is noted a sharp activation of the SAS. The excretion of A increases by 3.9 times, of NA--by 6.1 times, of DA--by 4.5 times with $P<0.001$ (relative to the background). There was also indicated an increase in the coefficients NA/A from 3.4 (background) to 5.7; of NA/DA--from 0.06 to 0.09 and A+NA/DA--from 0.08 to 0.12, which indicates sufficiently pronounced reserve abilities of the SAS in sexually immature individuals under conditions of severe ES and in addition a greater predominance of the activation of the mediator (NA) unit over the hormonal (A) unit in comparison with the reaction of the SAS in adult animals.

The excretion of NA in sexually immature monkeys exceeded the amounts of its excretion under severe ES in adult animals by 36% ($P<0.05$), while the excretion of A was 47% less. The restoration of KA excretion to the initial values was observed by the second day of the post-period.

As seen in Illustration 16, the maximum activation of the SAS under repeated action of stress irritants is seen in response to the second immobilization. The level of A in the daily urine relative to the initial values increased by 5.9 times, NA--by 6.9 times, DA--by 5.3 times (with $P<0.0001$). The content of A, NA and DA in the urine exceeded the level of their excretion noted in response to the first immobilization by 49% ($P<0.05$), 13% and 17%, respectively. To the following stress applications, just as in adult monkeys, there occurred a relative drop in the response reactions of the SAS. The more significant changes under conditions of repeated ES were noted in the excretion of A and NA. Thus, by the sixth immobilization the amount of excretion of A and NA were less than the amounts of their excretion to the first stress application by 1.7 times ($P<0.05$), and 1.3 times ($P<0.02$), respectively. The excretion of DA throughout all immobilizations remained significantly elevated and by the tenth immobilization its level, just as at the first stress application, exceeded the initial values by 460% ($P<0.0001$). There also took place a reduction in the size of the ratios (NA/DA) and (NA+A/DA). An analysis of the dynamics of KA excretion in the subsequent days following the end of each repeated immobilization shows a sufficiently pronounced tendency toward normalization of the function activity of the SAS, which is supported also by the rapid restoration (by the second-third day) of KA excretion to the initial amounts after cessation of the action of the stress irritants.

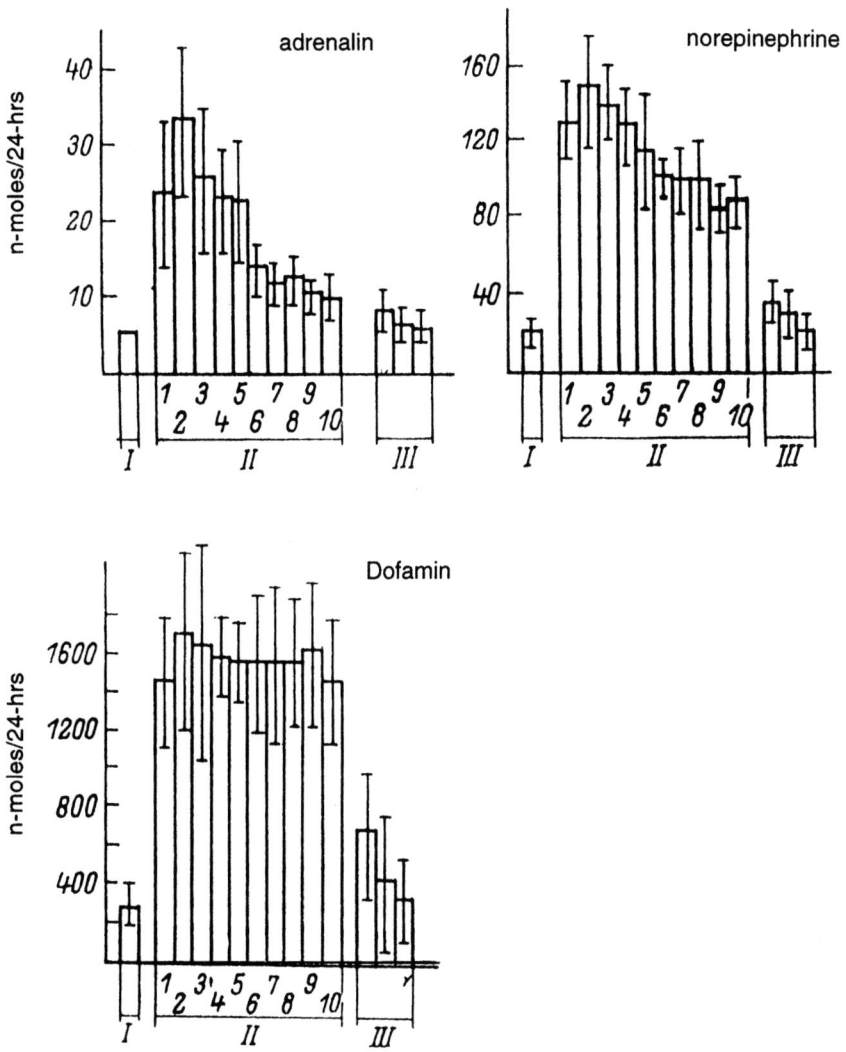

Illustration 16. Daily excretion of catecholamines in urine of sexually immature monkeys under repeated emotional stress. Mathematical averages with reliable intervals.

Y-axis - : amounts of catecholamine excretion, n-moles/24-hrs; X-axis - : I - background, II - cycle of 2-hour immobilizations; III - post-stress period (3 days); 1,2,3,...10 - corresponding immobilizations.

The results of the research conducted testified to the fact that under conditions of repeated emotional stress in sexually immature monkeys there occurs a pronounced restructuring of the functioning of the SAS, which is manifested in a relative reduction in the amounts of the excretion of A and NA, while the level of DA excretion remains sufficiently high throughout all stress applications. An analysis of the dynamics of KA excretion and of the sizes of the coefficients NA/DA and A+NA/DA allows us to assume that under these conditions there occurs a relative reduction in the activity of dofamin-beta-hydroxylase and a reduction in the formation of NA.

However, available data indicate that under conditions of chronic stress there occurs a significant increase in the activity of this enzyme (Kvetnansky, Mikulay, 1970; Kvetnansky, et al., 1970, 1971), while the observed reduction in the stress level of free KA is caused by an increase of their metabolism (Bolshakova, et al., 1972; Menshikov, et al., 1977; Menshikov, Bolshakova, 1978). A significant increase in the discharge of vanilil-phenyl-glycolic acid--the fundamental metabolite of KA, along with a reduction in the stress amounts of excretions of A and NA was observed by us in sexually mature papio hamadryus males during the conduction of daily immobilizations (Chirkov, 1984). The data we obtained allow us to assume that the decrease of the stress amounts of the excretion of A and NA to repeated immobilizations, which we observed in our experiments, is evidently related to an increase in KA metabolism and reflects on the whole an adaptive tendency of the restructuring of the function of the SAS under conditions of the repeated action of neuro-emotional irritants.

At the same time the significant sizes of the discharge of DA allow us to assume that under conditions of repeated ES in sexually immature monkeys there evidently occurs and increase in the activity of the enzyme DOFA-decarboxylase, which ensures a high level of dofamine production throughout all stress applications. The possibility of this theory is confirmed by available data which attests to the sharp increase in the activity of KA synthesizing enzymes, including DOFA-decarboxylase, under repeated immobilizations in rats (Kvetnansky, et al., 1971, 1978). In addition it is not excluded that along with an increase in the activity of DOFA-decarboxylase under conditions of chronic ES there occurs a more rapid transformation of DOFA into DA in comparison with its own synthesis. This agrees very well with available data indicating the possibility of a development of a relative shortage of DOFA in man and animals under the influence of intensive and repeated stress irritants (Mukhin, et al., 1978; Gubachev, Stabrovsky, 1981).

THE NATURE OF THE CHANGE OF THE LEVEL OF HYDROCORTISON AND TESTOSTERONE IN THE BLOOD OF SEXUALLY IMMATURE MONKEYS UNDER CONDITIONS OF REPEATED EMOTIONAL STRESS

During the period of adapting to confinement in individual cages and to the conditions of the experiment in male papio hamadryus of a prepubescent age there was noted a tendency toward a reduction in the concentration of hydrocortisone in the blood, the level of which, however, exceeded the amounts of this hormone in sexually mature animals by the end of the corresponding period by approximately 2.5 times with $P<0.02$; the concentration of testosterone in the blood of sexually immature animals by the end of the fourth week amounted to on the average (2.8 ± 0.11) n-moles/l, which was 11.5 times $(P<0.0001)$ less than the level of this hormone in sexually mature males (Table 3).

Table 3. Dynamics of the hydrocortisone and testosterone content in blood of monkeys during a period of adaptation to confinement in individual cages, and to conditions of the experiment.

Monkeys	Periods of Research (weeks)			
	First	Second	Third	Fourth
hydrocortisone (n-moles/ltr)				
Sexually immature	1693±312	1334±174	1412±165	1459±273
Sexually mature	1031±52	815±41	623±34	582±60
		$P_1<0.02$	$P_1<0.001$	$P_1<0.001$
		$P_2<0.05$	$P_2<0.01$	$P_2<0.02$
testosterone (n-moles/ltr)				
Sexually immature	2.46±0.59	2.46±0.21	1.83±0.21	2.80±0.11
Sexually mature	21.1±2.3	28.6±2.9	40.3±3.9	34.8±7.1
			$P_1<0.01$	
	$P_2<0.0001$	$P_2<0.0001$	$P_2<0.0001$	$P_2<0.0001$

Note: P_1 - reliability of the differences relative to the first week; P_2 - reliability of the differences in hormone content in the corresponding periods, relative to sexually immature monkeys.

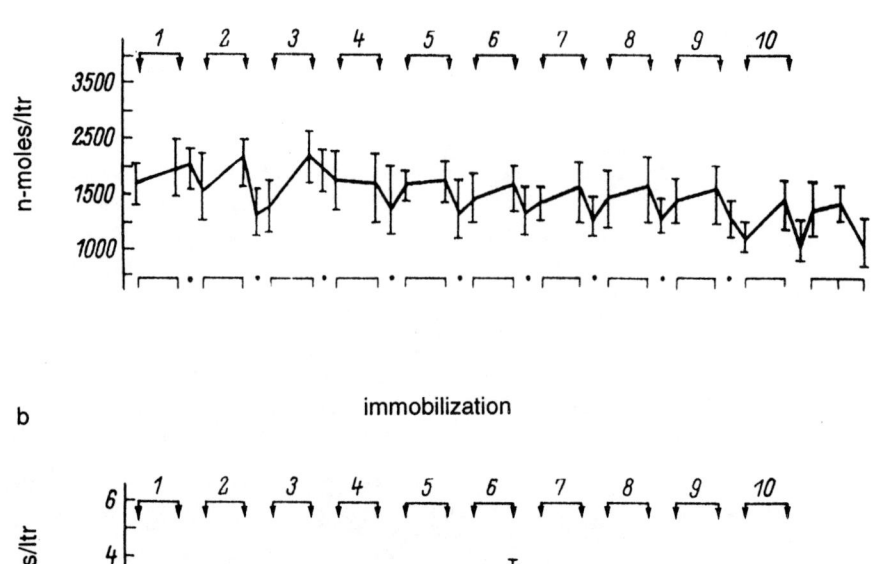

Illustration 17. Dynamics of hydrocortisone (a) and testosterone (b) content in the blood of sexually immature monkeys during repeated emotional stress. Mathematical averages with reliable intervals.

Y-axis - hormone concentration, n-moles/ltr; X-axis -- time of blood draw: 0 -before immobilization, 3, 24, - respectively, 3 and 24 hours after the start of each immobilization; 48, 72, 96 - elapsed time since the start of the 10th immobilization.

The differences we revealed in the content of hydrocortisone and testosterone in the blood of sexually immature monkeys in comparison with the concentration of these hormones in adult individuals reflect the age characteristics of the secretory activity of the cortex of the adrenals and the gonads, and is in agreement with the data of Butnev (1980), obtained during the study of the development of the endocrine function of the cortex of the adrenals and gonads in male papio hamadryus during ontogenesis. As the author indicates, a higher level of hydrocortisone and certain of its predecessors (11-desoxyhydrocortisone, 17-oxypregnenolone, 17-oxyprogesterone) in the blood of sexually immature monkeys at the age of 1-2 years in comparison with adult papios is the result of a less intensive metabolic transformation of gluco-corticoids, while the borderline low content of testosterone is caused by the immaturity of the endocrine function of the testes, the development of which in the given species of monkey occurs by the fourth year of age.

Under conditions of severe emotional stress (the first immobilization) in all sexually immature monkeys 3 hours after the beginning of the stress application there was noted an increase in the content of hydrocortisone in the blood of, on the average, 20% ($P_{kz}<0.05$) by comparison with the initial amounts (Illustration 17). 3 hours after the beginning of the second and third immobilizations the hydrocortisone level was maximum and amounted to (2116±334) n-moles/l and (2141±127) n-moles/l, respectively, i.e., it exceeded the baseline amounts by 40 and 70% ($P<0.05$). In spite of the significant amounts of the hormone concentration in the blood, the amplitude of its increase was relatively low as a result of the increase in the baseline level of hydrocortisone, characteristic for sexually immature monkeys. In response to the fourth stress application, which began with even higher baseline amounts, in all animals 3 hours after the beginning of the immobilization there was noted a drop in the level of the hormone in the blood.

Beginning with the fifth stress application, along with the tendency for a reduction in the baseline level of hydrocortisone, there occurred a decrease in the amounts of its increase in response to all subsequent immobilizations. Simultaneously there was noted a tendency toward a reduction in the hormone concentration 24 hours after the beginning of the stress applications. 96 hours after the beginning of the final--the tenth--immobilization in monkeys there was noted a significant drop in the concentration of hydrocortisone, the level of which amounted to on the average 60% ($P<0.05$) of the initial values, noted before the beginning of the first stress application.

As we see from Illustration 17, in sexually immature papio hamadryus males there is an absence of statistically reliable changes of testosterone concentration in the blood throughout the entire period of the stress applications. The testosterone content in the blood 3 hours after the beginning of the first, second and third immobilizations was virtually unchanged and the level of the hormone remained within low amounts, characteristic for sexually immature monkeys of this species. At the fourth, sixth and seventh immobilizations there was observed an insignificant increase in the concentration of testosterone in the blood, however these amounts during this period statistically reliably did not differ from the baseline level. It should be noted that 24 hours after the beginning of each repeat immobilization there was noted a tendency toward a reduction in the content of androgen in the blood relative to the baseline amounts. An insignificant reduction in the concentration of testosterone was noted also by the third and fourth days after the end of the cycle of stress applications, and furthermore by the fourth day the level of the hormone was minimal--(1.28±0.14) n-moles/l--and amounted to 55% ($P<0.05$) of the initial values registered before the beginning of the first immobilization.

Table 4. Correlative coefficients (r) between excretion of KA and hydrocortisone content in the blood of sexually immature monkeys under repeated emotional stress.

	Adrenalin	Norepinephrine	Dofamin
Hydrocortisone*	0.70 $P<0.01$	0.72 $P<0.01$	0.51 $P<0.05$

Note: *-hydrocortisone concentration 3 hours after the start of immobilization.

Thus, the study we conducted of the functional state of the steroid producing glands in sexually immature papio hamadryus males under conditions of repeated ES allowed us to establish that the maximum increase of the adreno-cortical activity occurs in response to the second and third immobilizations, i.e., during the period which corresponds to the greatest activation of the SAS. Along with the restructuring of the functioning of the SAS in response to repeated immobilizations, in sexually immature animals there occurs a rapid suppression of the adreno-cortical reaction, which is manifest in a positive correlation of the level of hydrocortisone and KA throughout the entire cycle of the stress applications (Table 4) and is in agreement with available data on the suppression of the adreno-cortical reaction during numerous immobilizations in mice (Riegle, 1973; Mikulay, et al., 1974).

It should be noted that the rapid suppression of the adreno-cortical reaction observed in our experiments can be related to the sufficiently lengthy intervals (2 days) between the repeated immobilizations, since, as shown in the experiment on sexually mature papio hamadryus males, the duration of the intervals up to 2 days is one of the adapting factors, which lead to the reduction in the response of GGAKS to repeated stress applications (Chirkov, 1984). As opposed to adult animals, in sexually immature monkeys in a setting of a higher baseline level of hydrocortisone in the blood, there are also registered lesser amounts of the amplitudes of its increase in response to the action of the stress irritant, which agrees with available data on the dependence of the stress reaction of the GGAKS on the baseline level of gluco-corticoid secretion (Markel, et al., 1981; Kazin, 1982). A distinguishing characteristic of the nature of the hormone balance in sexually immature monkeys, both under conditions of physiological calm and stress, is the borderline low content of testosterone in the blood, caused by the immaturity of the endocrine function of the testes.

CHANGES IN THE FUNCTION OF THE CARDIOVASCULAR SYSTEM IN SEXUALLY IMMATURE MONKEYS UNDER CONDITIONS OF REPEATED EMOTIONAL STRESS

During the period of adapting to confinement in individual cages and to the conditions of the experiment, in all monkeys of prepubescent age throughout the first week high values of AD were registered with an increase in the level of the systolic pressure to maximum values, amounting to, on the fourth day, (164.4±22.6) mm rt. st., which exceeded the corresponding values on the first day by 28% ($P<0.05$) (Illustration 18). At the beginning of the second week there occurred a normalization of the systolic AD and a significant decrease in diastolic pressure ($P<0.05$). Throughout the following three weeks we observed a stabilization in the amounts of AD, which at the end of the period of adaptation on the average for 5 monkeys amounted to (mm rt. st.): AD_s 128.3 6.1, AD_d 72.7 6.2, AD_{sr} 96.0 10.3 and AD_{pulse} 55.6 2.4.

Under conditions of repeated ES in all sexually immature monkeys there was noted a development of a sufficiently pronounced hypertensive reaction. As seen in Illustration 19, 3 hours after the beginning of the first immobilization AD_{sr} increases on the average to (130.7±3.1) mm rt. st., which is higher than the initial values by 20% ($P<0.05$) and remains at this level throughout 24 hours. 72 hours later, i.e., before the beginning of the second immobilization, there was noted a normalization of the amounts of AD. Beginning with the second stress application there is observed a persistent increase in the amounts of AD_{sr}, which is maintained throughout the entire period of the immobilizations and the following 6 months of the post-period. It should be noted that the increase in the amounts of AD_{sr} occurs primarily at the expense of an increase in the level of the diastolic pressure, which is indicated also by the high positive correlation (R = 0.92, $P<0.01$) between the dynamics of the amounts of AD_{sr} and the change AD_d.

The change in the systolic pressure under these conditions was less pronounced and only on the sixth immobilization was there noted an insignificant increase in the amounts of AD_{sr} on the average to (168.9±±2.1) mm rt. st., which altogether was greater than the baseline values

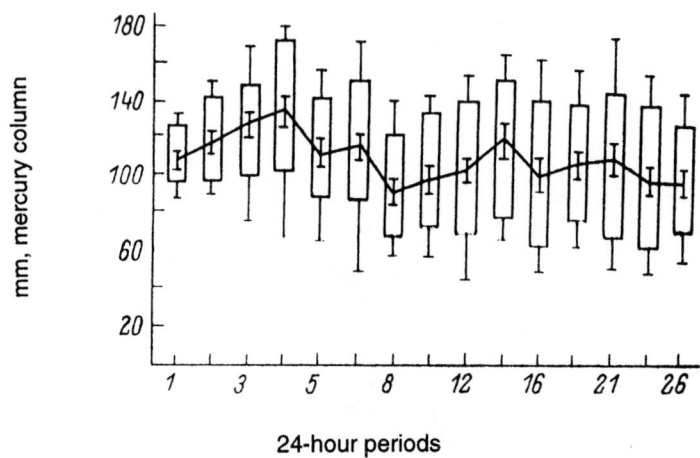

Illustration 18. Dynamics of change in arterial pressure (AD) in sexually immature monkeys during the adaptation period. Mathematical averages with reliable intervals.

Y-axis - : amounts of AD, mm mercury column; X-axis - : time of adaptation, in 24-hr periods; upper column limit - AD systolic; lower column limit - AD diastolic.

by 11.5%. An increase in the diastolic pressure during the absence of a change in the AD_{sr} led to a decrease in the amounts of the pulse pressure, which turned out to be reduced by approximately 50% ($P<0.05$) throughout all ten immobilizations and the post-period. During this the AD_{pulse} had a negative correlation ($R = -0.36$, $P<0.05$) with the dynamics of the amounts of AD_{sr}. Our attention is drawn to the high amounts of the frequency of the heart rhythm (CSR), registered in sexually immature monkeys both during the period of adaptation and throughout the entire cycle of the stress applications. The high frequency of the heart rhythm, observed before the beginning of repeated immobilizations, evidently reflects emotional stress, developing in monkeys in response to the action of a number of situational irritants, including anthropogenic factors.

The high sensitivity of the CSR to anthropogenic factors is indicated by experiments with the radio telemetric registration of the EKG, in which there was revealed a sharp increase in the heart rhythm during the mere approach of the experimentor to the monkey (Tatoyan, Chirkovich, 1972). Along with the high initial values of

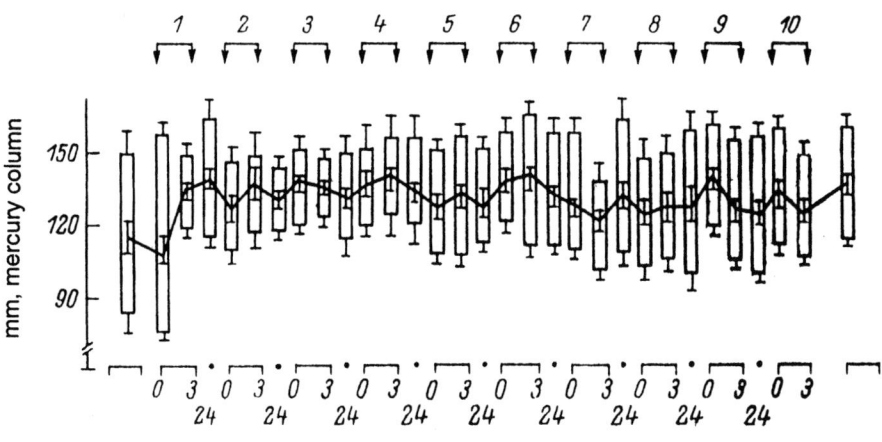

Illustration 19. Dynamics of change in arterial pressure (AD) in sexually immature monkeys under repeated emotional stress. Mathematical averages with reliable intervals.

Y-axis - values of AD, in mm. mercury column; X-axis - time of AD measurement - 0 - before immobilization, 3, 24 - after the corresponding time interval after the beginning of immobilization; post-stress period - measurement of AD throughout the following 6 mos. Other figures the same as for illus. 18.

CSR by the end of each immobilization there was noted a tendency toward a reduction in the amounts of this indicator on the average by 10-15%. The relative reduction in frequency of the heart rhythm during repeated immobilizations in monkeys in a setting of high initial values agrees very well with available data on the dependence of the intensity of the reaction on initial (baseline) frequency of the heart activity (Korneva, 1965). The revealed nature of the change of the amounts of AD_d and CSR is clearly manifest in the dynamics of the indicator of the vegetative index of Kerdo (Sokolov, Belova, 1983), calculated according to the formula:

$$(1 - AD_d/CSR) \times 100.$$

Table 5. Change in the vegetative index (VI) under repeated ES in sexually immature monkeys.

Experimental Conditions	VI	
	0	3
Background immobilization:	+64.0	
1st	+0.69.6	+38.3
2nd	+55.1	+41.3
3rd	+41.4	+31.6
4th	+46.5	+39.8
5th	+45.4	+39.8
6th	+40.9	+37.8
7th	+47.8	+40.2
8th	+46.3	+36.6
9th	+44.0	+39.6
10th	+38.6	+33.7
Post-stress period	+50.0	

Note: 0 - values of VI before the start of immobilization; 3 - values of VI 3 hours after the start of immobilization.

As we see in Table 5, throughout all stress applications the amounts of this index were highly positive, which attests to, according to available theories, the pronounced predominance of the sympathic-tonic influences (Sokolov, Belova, 1983). At the same time the observed reduction in the amounts of the vegetative index by the end of each immobilization, and also the tendency toward a decrease in its values according to the application of stress irritants, may be related to the relative decrease in the central adrenergic influences.

An analysis of the EKG established that under the influence of stress irritants in sexually immature monkeys there is noted a decrease in the amplitude of the T-tine in the III standard and the AVF points of contact and its significant increase in the chest points of contact. The more obvious dynamics of the amplitude of the T-tine is manifest in the analysis of its summary profile (ΣT) in 3 standard (ΣT_{1-3}) and 6 chest (ΣT_{v1-v6}) contact points of the EKG (Illustration 20). The characteristic trait of the change of in the chest points of contact was a significant increase in the amount of this indicator during the initial periods of all repeated immobilizations, which on the average in the group of animals amounted to (182.7±14.8) ($Pk_z<0.05$) relative to the baseline values, adopted for 100. At the end of each stress application there

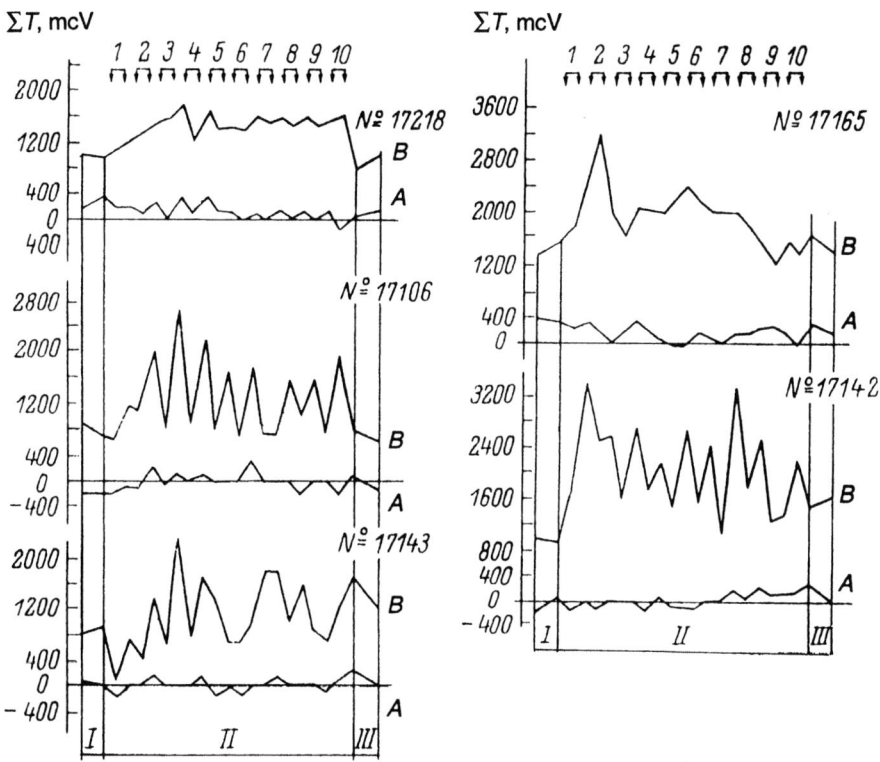

Illustration 20. Dynamics of the change in the summary amplitude of the Σ spike (ΣT) in three standard (A) and six chest contact points (B) EKG in sexually immature monkeys under repeated emotional stress.

Y-axis - : amount of in microvolts; X-axis - : I - background; II - cycle of stress applications; III - post-stress period; right - animal number. Other symbols same as in illustration 18.

was noted a lesser increase of v1-v6 and its increase in 3 hour intervals throughout the entire cycle of immobizations on the average for 5 monkeys amounted only to 37.3%, which along with the obtained data on the relative decline in frequency of the CSR by the end of each immobilization can testify to the development of adaptive restructuring of the regulation of cardial function, aimed at decreasing the central adrenergic influences (Belkania, 1982). Furthermore according to the application of repeated stress irritants there was also noted a tendency toward the normalization of the amount of the T-tine in the standard points of contact (Illustration 20). In 2 animals there were episodically observed lone stomach extrasystoles. However the indicated changes of the EKG bore a transitory nature and completely disappeared at various times during the post-stress period.

The results of our experiments agree with the data of other authors on the high reactivity of the cardio-vascular system in monkeys (Belkania, 1982), and also with research indicating that emotional stress and, in particular, emotional fear and alarm in these animals are accompanied by an increase in the central adrenergic influences and are manifest in a predominant increase on the EKG spatial vector of the T-tine (Cherkovich, et al., 1983).

Thus, as a result of the studies conducted it was established that during conditions of repeated ES in sexually immature monkeys there develops a hypertensive reaction, manifest in the predominant increase of diastolic pressure and in the reduction of the amounts of the pulse AD, which indicates the increase of the pressor regulation mechanisms and, in particular, the increase of vascular tension. This nature of a hemodynamic change has much in common with the dynamics of the cardio-vascular reaction in healthy people in response to emotional stress, which is manifest in the increase of AD_{sr} due to a progressive increase of the overall peripheral resistance (vasoconstruction), leading to a significant decrease in pulse pressure and a gradual drop in the minute volume of blood, which occurs in a setting of relative decreased rate of CSR (Sokolov, Belova, 1983). In spite of the normalization of pulse pressure 24 hours after the beginning of the first immobilization, in sexually immature animals during this period there is also observed an increase in the amounts of AD_{sr} and of diastolic AD, which indicates the intensity of the reaction to the applied stress stimulus, as well as the relative torpidity in the restructuring of vascular tension. These data agree with results of the research on modeling neurogenic arterial hypertension in monkeys, in which there was also observed a predominant increase of diastolic pressure at the initial stage of the formation of arterial hypertension (Startsev, 1971, 1976, 1977; Repin, Startsev, 1975).

At the same time in the experiments of Belkania and Dartsmelia (1984) in modeling neurogenic AG under conditions of orthograd statics in male rhesus macaques there was established a phasic nature of AD change, and furthermore following the initial reduction in the AD level there occurred an increase in diastolic AD and then the joint increase of both systolic and diastolic AD. The absence in our experiments of the phase of the initial reduction of AD can first and foremost be explained by the difference in the nature of the stress irritants, and also by species and age characteristics of the reaction of the cardiovascular system to stress.

In a comparison of the dynamics of AD in sexually immature male papio hamadrus to the nature of the response reaction of the circulation system in adult monkeys, under analogous conditions of ES, in the latter there was revealed a rapid restoration of the amounts of AD to the initial level after the end of each stress application, which indicates a greater resistance of the cardio-vascular system to repeated ES in sexually mature animals (Chirkov, et al., 1986). Along with this in sexually immature monkeys and adult individuals differences were revealed in the nature of the behavior reaction in response to submission of the animals to the experiment. In sexually immature animals during capture and their removal from the cage there is observed a passive-defense reaction, while in adult papios aggressive elements of behavior predominate. At the same time, manifestations of depressed behavior states in sexually immature animals, arising as the result of stress, were less pronounced in comparison with adult monkeys.

In comparing the change of the functional state of the SAS and the steroid producing glands with the dynamics of the AD increase in sexually immature monkeys under conditions of multiple application of stress irritants it was revealed that the maximum increase of the sympatho-adrenal and adreno-cortical activity noted in the second and third immobilizations corresponds to the initial phase of the development of a resistant hypertensive state. Later, according to the stabilization of the elevated level of AD in response to repeated immobilizations, there occurs a reduction in the stress amounts of the excretion of A and NA and a suppression of the adreno-cortical reaction. The analysis which was conducted of the correlative relationship between the dynamics of AD and the hormonal indicators revealed the presence of a positive correlation ($R = 0.45$, $P<0.05$) only between AD_d and dofamine, the excretion of which remained significantly elevated in response to all repeat immobilizations. The presence of pronounced functional interrelationships between the production of dofamine and the amount of AD are also witnessed by clinical research in which it is shown that the determination of DA content is necessary to determine the stage of the development of hypertensive illness (Snider, Kuchel, 1985).

The results of our research, which show that the development of a resistant hypertensive state in sexually immature monkeys corresponds to the initial stage of the discharge of significant amounts of KA, with the pronounced predominance of the excretion of NA over A, is in agreement with available data on the change of the activity of the SAS under modeling of the neurogenic AG in other laboratory animals (Sudakov, 1976; Belova, Kvetnansky, 1981), and also agrees with numerous studies showing the important role of KA in the regulation of the cardio-vascular system and, in particular, in the mechanisms of AD increase during stress (Miloslavsky, et al., 1971; Shalyapina, 1979; Markel, 1983; Sokolov, Belova, 1983). The data obtained in our experiments on the restructuring of SAS functioning while maintaining an elevated level of AD in sexually immature monkeys agrees with the results of many studies, which show the phasic nature of the change of the functional activity of this system in man in various stages of the formation of hypertensive illness, and it has been established that the increase of the sympatho-adrenal activity is noted only during the initial stages of the illness, while further changes in KA secretion have a compensatory nature (Berezkin, Tarasov, 1978; Grigorieva, 1978; Vasilyev, 1981; Sokolov, Belova, 1983). The support of the elevated level of AD meanwhile can be secured by the increase in the sensitivity of the adreno-receptors and the engagement of other humoral pressor mechanisms, which participate in AG stabilization (Sokolov, Belova, 1983).

We must note that in response to the first immobilizations in sexually immature monkeys between the amounts of the content of hydrocortisone in the blood and ADsr there was noted a highly positive correlation ($R = 0.89$, $P<0.01$), which along with the increase in hydrocortisone content in the initial state, as well as the unidirectional change in the activity of the SAS and the GGAKS evidently witnesses the participation of hydrocortisone in the mechanisms of the elevation of vascular tension, and consequently, in the increase of diastolic AD. Studies on rats have also shown a positive correlation between the increase in the level of gluco-corticoids in the blood and the development of AG under conditions of chronic neurogenic stress, and furthermore the increase of hydro-corticoid production preceded the increase of AD amounts (Bankova, et al., 1976). The engagement of gluco-corticoids in the mechanisms of pressor reaction of emotiogenic origin are indicated by studies which have shown that the stable increase in the AD level during modeling of the arterial hypertension is possible only with the joint increase with the sympatho-adrenal and adreno-cortical activity (Sudakov, 1976). While studying the functional state of the GGAKS in patients with hypertensive illness there was also noted an insignificant increase in the level of gluco-corticoids in the blood, and as a result of which the theory developed of the secondary role of these hormones in the pathogenesis of hypertensive illness (Miloslavsky, et al., 1971; Sokolov, Belova, 1983).

It should be noted that the most characteristic peculiarity of the hormonal reactions in sexually immature monkeys to the influence of repeated stress irritants was the extremely low reactivity of the system pituitary-gonads, and furthermore the content of the anabolic hormones in the blood in these animals was 10 times lower in comparison with adult males. Considering the available data on the development of AG in gonadectomized animals (Vartapetov, Gladkova, 1971), and also the ability of androgens to reduce the level of KA in tissues (Bakhova, 1975; Van Loon, 1978; Chirkov, 1984), one can assume that the formation of a resistant arterial hypertension in sexually immature papio hamadryus males is related to the nature of the hormonal balance and, in particular, the absence of the inhibiting effect of testosterone on the activity of the SAS as a result of its low content in the blood.

Thus, the results of our research show that the initial increase of the sympatho-adrenal and adreno-cortical activity with the absence of a reaction of the endocrine function of the testes, observed under conditions of repeated ES in prepubescent monkeys, is accompanied by the development of a persistent arterial hypertension, manifest primarily in the increased diastolic AD. According to the action of the stress irritants along with the stablization of the elevated level of AD, there is noted a decrease in the stress amounts of the discharge of A and NA, and the suppression of the adreno-cortical reaction. The data obtained allow us to make the assumption that the development of a persistent arterial hypertension in prepubescent animals is related to the pronounced characteristics of hormonal balance and, in particular, the low level of testosterone content in the blood.

Chapter 6

THE HUMORAL-HORMONAL RELATIONSHIP UNDER EMOTIONAL STRESS IN MONKEYS

In recent years, characterized by the stormy and successful development of the problem of neuro-humoral regulation of the physiological functions, interest in the study of the humoral-hormonal and interhormonal relationships under the widest variety of physiological and pathological states of the organism, has sharply increased. A wealth of material has been accumulated, witnessing the fact that in the regulation of physiological, biochemical and immunological processes, determining the maintenance of the homeostasis under the influence of stress factors, which disturb the constancy of the internal environment of the organism, the decisive significance belongs not so much to the level of the content in the liquid environment of the organism of certain hormones and mediators, their predecessors and active products of metabolism, as to the relationship of biologically active substances, the balance of the ergo-and trophotropic humoral influences, to the level of reactivity and to the reserve capabilities of the SAS and other neuroendocrine complexes, their interrelationship with various neuropeptides and neurohormones, to the formation of complexes of biologically active substances, to the activity of regulating compensatory mechanisms, etc..

In spite of available works on the complex research of the activity of the SAS and GGAKS under the influence of extreme irritants, the nature of the functional interrelationships of these systems under the conditions of stress remain insufficiently studied. In addition, between KA and gluco-corticoids there has been revealed a close interaction, which is followed on various levels of the organization of the neuroendocrine and physiological functions. Much support has been gathered on the participation of the adrenergic structures of the brain in the activation of GGAKS under stress, and furthermore many authors feel that KA-ergic neurons have a stimulating influence on the release of KRF and the secretion of AKTG (Shalyapina, 1976; Shalyapina, Rakitskaya, 1976; Sapronov, 1980; Amar, et al., 1982; Johnstone,

Ferrier, 1980; Axelrod, Reisine, 1984). The most obvious role of the KA-ergic system in the regulation of GGAKS is manifest in the limit of the functional abilities of the adrenergic mechanisms in the reduction of the KA reserves in the brain, when in response to the stress application additional amounts of KA cannot be emitted from their reserve locations and there occurs a breakdown in the regulatory mechanisms which leads to a disturbance of the response reaction of the adrenal cortex (Shalyapina, 1976). In the opinion of many authors, an activating influence on the secretion of gluco-corticoids is also made by the serotonin-ergic system (Naumenko, Popova, 1975; Gan-ong, et al., 1976; Fuller, 1981; Amar, et al., 1982; Smuthe, et al., 1983). In addition a number of works cite evidence of the absence of the influence of the serotonin brain on the adreno-corticotropic activity of the pituitary (Karteszi, et al., 1981) and even on the inhibiting effect of the serotonin-ergic system on the activation of the GGAKS under stress (Vermes, Teledgu, 1978; Frey, Moberg, 1980; Bruni, et al., 1982).

However, the majority of authors point to the activating influence of the serotonin-ergic system of the brain in the regulation of GGAKS under the influence of stress irritants (Naumenko, Popova, 1975; Stabrovsky, et al., 1981; Fuller, 1981; Kennet, Joseph, 1981; Amar, et al., 1982; Tsulaya, 1985). These contradictions, in the opinion of Viru (1981), are related to the fact that the morphofunctional organization of the integral process of the direction of this system also includes the interaction of various nerve structures, which not only are diffusely situated, but are also different from each other according to their neuro-chemical properties. Furthermore, as the author feels, a chain transfer of regulating influences from structures with one neuro-chemical specification to another is not excluded. An important contribution to the understanding of this process is made by the work of Sapronova (1980) in which the selective blockade of various mediator systems of the brain, just as the disturbance of the synthesis or the exhaustion of the reserves of 1 or 2 mediators in the hypothalamus, has virtually no influence on the degree of the reaction of the system pituitary-adrenal cortex, elicited by any irritant.

The author assumes that in these conditions the transfer of the stimulating signal to the pituitary-tropic zone of the hypothalamus occurs by means of attracting other unharmed neurmediator systems. In these experiments the introduction of various stimulating substances of a mediator-type influence has shown that the activation of the system hypothalamus-pituitary-adrenal cortex can be elicited by the excitement of the central adreno-, cholino-, serotonin- and histamine-sensitive receptors. This gives us reason to feel that the functional activity of the hypothalamus-pituitary-adreno-cortical system is not determined strictly by a single monoaminergic mechanism of the brain.

An important unit of the functional interrelationships of the sympatho-adrenal system with GGAKS is the joint engaging of KA and gluco-corticoids in the biochemical processes on the level of affector organs. It has been established that in the absence of gluco-corticoids many physiological reactions and metabolic effects in response to the influence of KA drop out or are weakened (Utevsky, Gaisinskaya, 1971; Utevsky, Rasin, 1972; Osinskaya, et al., 1975; Papafilova, Palchikova, 1980).

The permissive effect of the hormones of the adrenal cortex on the activity of the sympathic nervous system, described by Ingle in the 50's (Ingle, 1952), has found confirmation in the present day. One of the mechanisms of this effect is the formation (with the help of gluco-corticoids) of the functionally active protein complexes of the mediator (NA) with receptor structures. The ability of the gluco-corticoids to increase the excitability of the receptor apparatus, to strengthen the afferent influence has been established, which, along with their influence on various structures of the CNS allows us to speak of the dependence of the degree of SAS activation on the functional state of the adrenal cortex (Nozdrachev, 1969). A disturbance of the steroid balance in various periods of embryogenesis leads to a change in the functional activity of KA-ergic neurons, eliciting in adult rats the inhibition of the hypothalamus-pituitary-adreno-cortical reaction to stress (Naumenko, Digalo, 1979).

Many studies have established the important role of the gluco-corticoids in the regulation of the adrenergic processes in the brain and on the periphery, and furthermore one of the more important mechanisms controlling KA production under stress is considered the influence of gluco-corticoids on the activity of catecholamine-synthesizing and KA-metabolizing enzymes (Shalyapina, Rakitskaya, 1976; Krachun, 1977; Ciaranello, 1979; Holzbauer, et al., 1979; Ventura, et al., 1979a; Markey, et al., 1982; Khaidarliu, et al., 1983; Axelrod, Reisine, 1984).

At the same time the interaction of KA and the hormones of the adrenal cortex in a setting of the changed reactivity of the neuroendocrine system, or under conditions of intensive and lengthy stress applications is seen as one of the main pathogenetic mechanisms in the development of experimental arteriosclerosis and arterial hypertension of an emotional origin (Sudakov, 1976; Homulo, 1982).

Regarding the participation of the central adrenergic mechanisms in the regulation of the system hypothalamus-pituitary-gonads there exists a single point of view supporting the activating role of norepinephrine and dofamine of the brain in the processes of synthesis and luliberin, and furthermore it is indicative that the adrenergic brain structures take part not only in the regulation of the tonic activity of the gonadoptropic function of the hypothalamus, but also in the support of the seasonal fluctuations of the level of the level of androgens in the blood (Naumenko, 1981; Markaryan, et al., 1983). There is also data in support of the activating influence on the endocrine function of the testes of the peripheral adrenergic mechanisms (Gladkova, et al., 1982). However, as a number of works show, the extreme injection of KA into the blood under stress or the introduction of exogenic KA into the system blood flow leads to the suppression of the production of androgens in the testes and the reduction of the testosterone level in the blood (Levin, et al., 1967; Damber, Johnson, 1978; Verhoeven, et al., 1979; Collu, et al., 1982; Gotz, et al., 1983).

As opposed to the activating role of the central KA-ergic systems on the endocrine function of the testes, serotonin has an inhibiting influence on the compensatory processes, which restore the lowered level of androgens in the blood (Naumenko, Popova, 1975; Naumenko, 1981; Markaryan, et al., 1983). Opposed to this are the results of studies in which there was revealed a stimulating influence of serotonin on the release of luliberin and on the secretion of gonadotropines (Baumgarten, et al., 1977; Ruzsas, et al., 1982).

It should be noted that the interaction of SAS with the hormonal function of the testes is not limited to the participation of KA in the regulation of the synthesis and secretion of testosterone. Available literature cites evidence of the inhibiting influence of the sex hormones and exogenic luliberin on the activity of the SAS, which is manifest in the slow down of the synthesis of KA in the brain, in a reduction of its level in the various organs and in the blood, in a reduction of the secretion in urine (Bakhova, 1975; Van Loon, 1978; Anisimov, 1979; Chirkov, 1984).

Recent neuro-chemical and neuro-pharmacological studies using radioligand methods of analysis have significantly enriched and expanded the concepts of the intimate mechanism of the interaction of KA and steroid hormones and have shown that in the regulation of the processes of adrenergic mediation and the functional activity of the systems hypothalamus-pituitary-adrenal cortex-testes, as well as their interaction under stress, an important roles belongs to the endogenic neuropeptides and, in particular, to opiates (Viru, 1981; Vedernikova, Maisky, 1981; Coste, Trabukki, 1981; De Souza, Van Loon, 1982; Charnoy, et al., 1982; Burov, Vedernikova, 1984). These studies have convincingly demonstrated the close structural and functional ties between the KA-ergic and opiate systems: They have established the presence of opiates in the chromassine cells of the adrenal glands, they have presented proofs of the localization of opiate receptors on the presynaptic DA-ergic terminals in the SNS and have shown that endogenic opiates are neuro-modulators of KA-neuromediation. They

Table 6. The dynamics of steroid hormone content in the blood of male Papio hamadryas during the period of adaptation to confinement in individual cages and to the experiment conditions. (Mm)

Steroids	Time of blood draw (daily)			
	1-c	8-c	15-c	22-c
Hydrocortisone	1031±52	815±41 $P<0.02$	623±34 $P<0.001$	528±60 $P<0.001$
11-desoxyhydrocortisone	18.7±1.2	14.6±0.7 $P<0.02$	12.1±0.6 $P<0.02$	14.8±0.8 $P<0.05$
Corticosterone	19.2±3.4	36.3±1.6 $P<0.01$	30.7±2.7 $P<0.01$	39.7±6.9
Testosterone	21.1±2.3	28.6±2.9	40.3±3.9 $P<0.01$	34.8±7.1

Note. P - reliability of differences relative to the first 24-hour period.

have proven the participation of endogenic opiate peptides in the regulation of the pituitary control of the adreno-cortical activity and the endocrine function of the testes.

In the opinion of a number of authors, the neuro-peptide systems, including the opioid system, has an exceptionally important role in the coordination and regulation of the neuro-endocrine functions, of the state of the psycho-emotional sphere, the behavior reaction and mechanisms of resistance of man and animal to emotional stress and to the influence of various stress and extreme factors (Vedernikova, Maisky, 1981; Yumatov, 1983; Burov, Vedernikova, 1984).

For the study of the functional interrelationship of the hormonal activity of the steroid producing glands and the SAS under emotional stress of relatively low intensity, throughout a 3 week period while the monkeys adapted to confinement to individual metabolic cages, we periodically conducted a collection of blood to determine gluco-corticoids and testosterone, and also a collection of daily urine for an analysis of the KA level. It was established that under these conditions in all animals there is a statistically reliable reduction in the level of gluco-corticoids and an increase in the concentration of testosterone in the blood (Table 6). The most significant change was revealed in the dynamics of the hydrocortisone content, the level of which in the blood by the end of the third week was reduced almost by half ($P<0.001$) in comparison with the first day. The reduction of the hydrocortisone concentration had a positive correlation (R=0.83,

Table 7. Correlation coefficients (r) between the level of steroid hormones and biogenic amines in monkeys during the period of adaptation to confinement in individual cages.

Steroid hormones	Catecholamines		
	A	NA	DA
Hydrocortisone	0.89 $P<0.01$	0.71	0.95 $P<0.01$
11-desoxyhydrocortisone	0.86 $P<0.05$	0.86 $P<0.05$	0.86 $P<0.05$
Corticosterone	0.82 $P<0.05$	0.83 $P<0.05$	0.76 $P<0.05$
Testosterone	-0.81 $P<0.05$	-0.79 $P<0.05$	-0.93 $P<0.01$

Note. P - reliability of the correlation according to Student.

$P<0.05$) with the change in its nearest predecessor, 11-desoxyhydrocortisone. Between the gluco-corticoids and testosterone a negative correlation was noted. The coefficients of the correlation between the level of hydrocortisone, 11-desoxyhydrocortisone, corticosterone and testosterone were correspondingly equal: -0.94, -0.94, -0.87 with $P<0.01$. It is interesting to note that, during the period of adaptation the dynamics of the KA reduction in urine had a highly positive correlation with the dynamics of the change of the level of gluco-corticoids and a negative correlation with the dynamics of testosterone content (Table 7). Consequently, the reduction of the adreno-cortical activity and the increase in the level of testosterone in the blood, observed during the period when the monkeys were adapting to new conditions of confinement, occurred simultaneously with the normalization of the functional state of the SAS.

An analysis of the functional state of the sympatho-adrenal and hypothalamus-pituitary-adreno-cortical, gonad systems under the influence of neuro-emotional stimulants, applied at various times of the day, showed that the development of severe ES in the evening hours, corresponding to the period of a reduction of the overall behavior and movement activity of the monkeys, is characterized by an increase in the reactivity of certain units of the neuroendocrine system. This is manifest in the more pronounced activation of the sympatho-adrenal system, in the greater amounts of the amplitude of the increase of gluco-corticoids and the drop of testosterone in the blood in comparison

Table 8. Correlation coefficients (r) between the level of biogenic amines and steroid hormones in monkeys under a 2-hour immobilization in the morning and evening hours.

Steroid hormones	Catecholamines		
	A	NA	DA
	Value of r under stress in the morning hours		
Hydrocortisone	0.91	0.99	0.87
	$P<0.01$	$P<0.01$	$P<0.01$
11-desoxyhydrocortisone	0.94	0.93	0.87
	$P<0.01$	$P<0.01$	$P<0.01$
Corticosterone	0.86	0.87	0.84
	$P<0.05$	$P<0.01$	$P<0.05$
Testosterone	-0.84	-0.84	-0.81
	$P<0.05$	$P<0.05$	$P<0.05$
	Value of r under stress in evening hours		
Hydrocortisone	0.97	0.98	0.98
	$P<0.01$	$P<0.01$	$P<0.01$
11-desoxyhydrocortisone	0.67	0.93	0.96
		$P<0.01$	$P<0.01$
Corticosterone	0.87	0.96	0.98
	$P<0.01$	$P<0.01$	$P<0.01$
Testosterone	-0.29	-0.89	-0.92
		$P<0.01$	$P<0.01$

Note. P - reliability of the correlation according to Student.

with the nature of the neurohormonal impulses observed during ES in the morning hours. During stress in the evening the amounts of KA excretion were also higher both in the amplitude of their increase, and in their absolute values, while the increase in the reactivity of steroid-producing glands was expressed primarily in the increase of the amplitude of the increased level of gluco-corticoids and the drop of testosterone in the blood with a virtually equal content of these hormones with their corresponding values under stress in morning hours.

A comparison of the correlations between steroid hormones and biogenic amines reveals a similar nature of the interaction of these hormones and mediators under stress in the morning and evening hours

(Table 8). As we see from this table, there were highly positive correlations between the change in the amounts of KA excretion and the level of gluco-corticoids, while there were negative correlations between KA content and the concentration of testosterone.

During the period of the post-effect of ES, produced in the evening hours, there was noted in monkeys a tendency to a more prolonged increase in the activity of the sympatho-adrenal and adreno-cortical systems, which agrees very well with the data of a number of authors, pointing to a significantly greater activation of the SAS and the GGAKS and a protracted nature of the normalization of KA excretion and the level of cortico-steroids under emotional stress in the night time in healthy people (Kassil, et al., 1973; Belova, Vasiliev, 1974; Vasiliev, 1981; Vasiliev, Chugunov, 1985). The combination of the increase in the activity of the SAS and the GGAKS observed during the period of the post-effect of ES can be explained by the mutual increase in the secretion of KA and gluco-corticoids, since it is well-known that not only catacholamines but gluco-corticoids also strengthen the synthesis and secretion of KA (Beitnis, et al., 1973; Shalyapina, 1976; Johnstone, Ferrier, 1980; Amar, et al., 1982; Axelrod, Reisine, 1984; Jones, et al., 1984).

It should also be emphasized that the high reactivity of the SAS and the steroid-producing gland in monkeys during stress in the evening hours does not signify a further error in the functioning of the hormonal system, but rather reflects the degree of expression of the adaptive reaction of the organism. At the same time during ES in the evening there was noted a more rapid restoration of the testosterone level in the blood to its initial amounts, which, on the one hand, points to the absence of the formation of a resistant emotional excitement, and on the other hand, to the relative independence of the GGGS reaction from the functional state of SAS and GGAKS.

In comparing data which characterize the functional state of the SAS and the steroid-producing glands in sexually mature papio hamadryus males under various conditions of the experiment, including the adaptation period and the influence of a 2-hour and 10-hour immobilization, there was established a clear relationship between the degree of expression of the hormonal reaction and the strength and duration of the influence of the neuro-emotional stimulants. It was shown that the greatest increase in the activity of the SAS and the GGAKS and a deep prolonged suppression of the endocrine function of the testes occurs during a 10-hour immobilization, less sharp hormonal impulses are noted during a 2-hour immobilization and a minimal expression of hormonal manifestations of the stress reaction is reveal under the influence of a relatively weak stress stimulant--the influence of the new setting and experimental procedures. It is important to note that the restoration of the concentration of hydrocortisone in the blood,

Table 9. Change in the amount of the relative coefficient (K) under repeated emotional stress with various regimens of the application of stress irritants.

Experimental conditions	Immobilization						
	Preliminary	1	2	3	4	5	6
Cycle of daily immobilizations		87.8	187.8	350.3	323.3	281.7	271.3
Analogous cycle of stress applications against a background of preliminary immobilization	82.0	55.8	63.4	78.5	70.1	52.0	68.6
Cycle of repeated immobilizations with intervals of 72 hours		93.5	61.8	56.2	60.8	55.1	

Note. $K = \dfrac{A + NA + F}{F}$, where A - the amount of adrenalin excretion (n-moles/ltr 24-hrs,) in the 24-hour period corresponding to each immobilization; NA - the amount of norepinephrine excretion (n-moles/ltr 24-hrs) in the 24-hour period corresponding to each immobilization; F - the sum of the amounts of hydrocortisone content n-moles/ltr before, and 2 and 6 hours after the start of each immobilization; T - the sum of the amounts of testerone content (n-moles/ltr) before, and 2 and 6 hours after the start of each immobilization.

occurring during the first days of the post-period of the 10-hour immobilization, did not coincide chronologically with the normalization of the testosterone content, which remained at a borderline low level for the following 2 days. This fact has great significance, since many researchers, in the capacity of a leading hormonal criterion of the adaptation of the organism to the influence of stress stimulants, use indicators of the adreno-cortical function, which, as our experiments show, do not always adequately reflect the anabolic phase of the stress reaction. The model of a 10-hour immobilization of sexually mature monkeys which is used is of significant interest for the further study of the nature of hormonal interrelations under stress, since under these conditions the restructuring of the hormonal balance of the organism toward a sharp activation of the CAS and GGAKS and a deep suppression of the endocrine function of the sex glands is most clearly manifest.

A comparison of the data obtained in the study of the function of the steroid-producing glands with the nature of the activation of the SAS under conditions of daily application of stress stimulants shows that the greatest increase of KA excretion in the urine corresponded to the maximum (amount) increase of hydrocortisone content and its predecessors, observed in response to the second and third immobilizations, and to a deep reduction in the level of testosterone in the blood. The application of preliminary immobilization or the increase in the duration of the intervals between individual stress applications led to a reduction in the peak increases of gluco-corticoids in the blood, and to the rapid restoration of the testosterone level, as well as to a significantly smaller increase in the amounts of the excretion of KA and VMK, which on the whole prevented a sharp increase in the changes of the hormonal balance in response to repeated stress influence. The most obvious differences in the formation of the hormonal reaction of the steroid-producing glands and the functional state of the SAS in three series of experiments are revealed in the analysis of the relative co-efficient (K) of the summary relation of the level of KA and hydrocortisone to testosterone (Table 9).

The predominance of the activation of the mediator unit (NA) over the hormonal unit (A), observed in all groups, can be seen as a manifestation of the defense mechanism, preventing overstress and the exhaustion of the function of the adrenal cortex, since it is well-known that adrenaline, as opposed to norepinephrine elicits a significantly greater secretion of AKTG. In their turn the high amounts of KA excretions, proving the sharp increase in their synthesis and secretion in response to repeated immobilizations, are evidently caused by a significant increase in the level of gluco-corticoids in the blood.

To illustrate the nature of the interrelationships of the functional state of the SAS and the steroid-producing glands under conditions of physiological calm and stress we applied a graphic method, allowing us to obviously express the relationship of the amounts of KA excretion and their corresponding level of steroid hormones in the blood. As we see in Illustration 21, the narrow range of the distribution of the indicators of KA discharge and hydrocortisone content in the blood corresponds to conditions of physiological calm. The greater distribution on the peaks corresponding to the content of testosterone and KA under these conditions reflects the more pronounced fluctuation in the level of testosterone in the blood, which can be explained by individual differences in the secretory activity of the testes in monkeys. Under stress there is observed a significant displacement of the points, which characterizes a sharp change in the functional state of the systems being studied. Along with the displacement there is

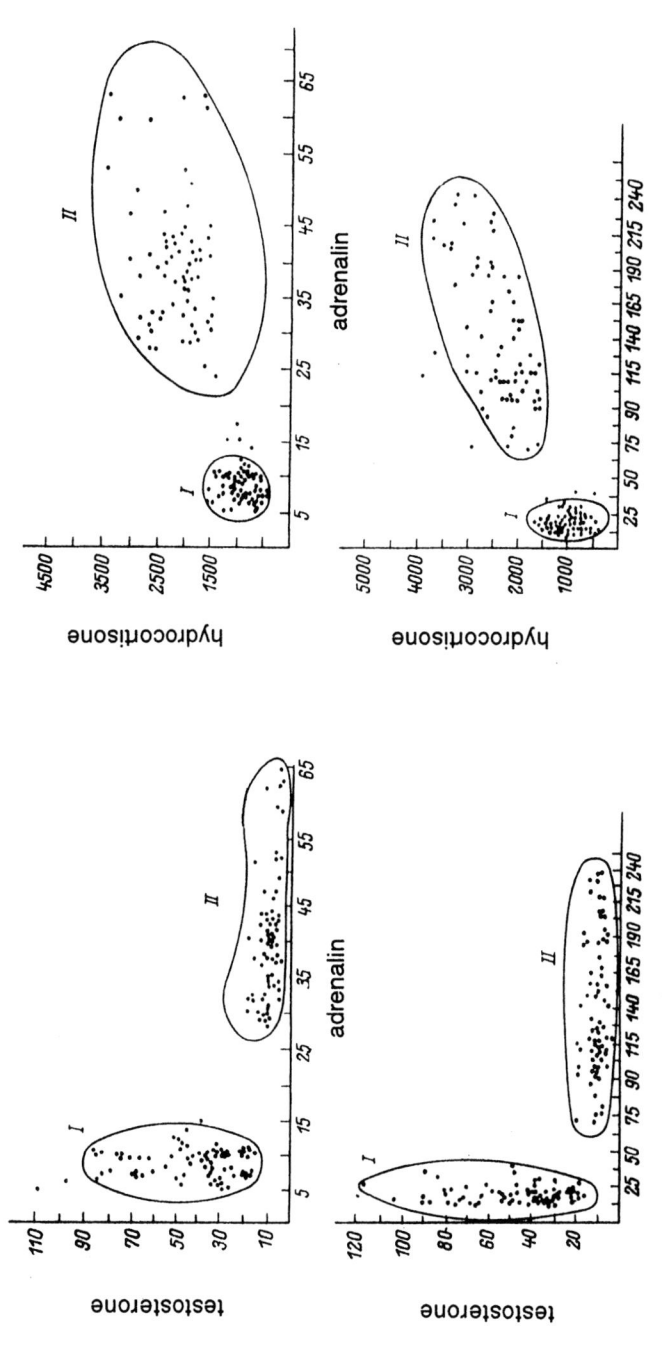

Illustration 21. Graphic relationships of the concentration of steroid hormones in the blood to the amounts of catecholamine excretion in monkeys under physiological calm (I) and stress (II).

Y-axis - concentration of hydrocortisone and testosterone, n-moles/ltr; X-axis - - amounts of adrenalin and norepinephrine excretion, n-moles/ltr.

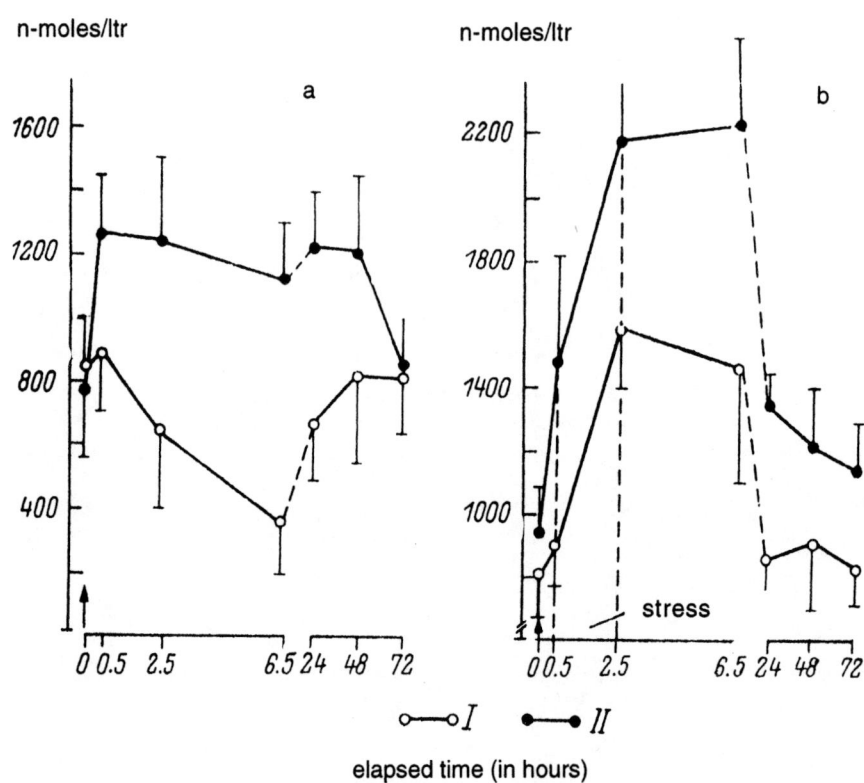

Illustration 22. Dynamics of hydrocortisone concentration in the blood of monkeys during injection with L-DOPA under physiological calm (a) and stress (b). Mathematical averages with reliable intervals.

Y-axis - concentration of hydrocortisone, n-moles/ltr; X-axis - time of blood draw: 0 - before injection, 0.5, 2.5, 6.5, 24, 48 and 72 - elapsed time from the time of injection, in hours; arrow - time of injection of 0.1g L-DOPA. I - control; II - experiment.

noted a dispersion of the points, caused both by differences in the regimen of the application of the stress irritants, and by a change in the reactivity of the system, occurring throughout the stress cycles.

The pronounced differences in KA excretion and in the dynamics of the change of the level of steroid hormones in the blood during a change in the duration of intervals between immobilizations indicate the dependence of the nature of the reactions of the sympatho-adrenal and hypothalamus-pituitary-adreno-cortical-gonad systems on the temporal regimen of the application of the stress irritant. The application of preliminary stress stimulus with a period of post-stress, sufficient for a complete restoration of the neuro-endocrine impulses, or the increase in the duration of the intervals between the action of similar stresses, equal in strength, leads to the optimization of adaptive restructuring of the reaction of the neuro-endocrine systems studied, which is manifest in the reduction of their reactivity and in the more rapid restoration of the functional activity during the post-periods.

As is well-known, the more obvious interhormonal relationship are revealed during the functional tests with different pharmacological and hormonal preparations. The study of the nature of the change of the adreno-cortical activity in experiments on monkeys using an injection of L-DOPA, increasing the synthetic capabilities of the SAS, allowed us to establish that L-DOPA in a dose of 0.1 g elicits in intact monkeys a significant and lengthy increase in the concentration of hydrocortisone in the blood (Illustration 22). The content of the hormone 30 minute after the injection of the preparation exceeded the initial level by 64% ($P<0.002$) and was 1.4 times higher than the control. The high amounts of hydrocortisone in the blood were registered after 2.5 and 6.5 hours, amounting to 191% ($P<0.001$) and 299% ($P<0.0001$), respectively, relative to the hormone content in the control experiment. The 3-fold increase in hydrocortisone content of the blood 6.5 hours after the injection of the preparation relative to the control indicates the preservation of the stimulating influence of L-DOPA on the secretion of gluco-corticoids during the period of the reduced daily activity of the adrenal cortex. The elevated level of hydrocortisone in the blood relative to the initial values and the control was noted also 24 hours and 48 hours after the injection of L-DOPA (Illustration 21).

Immobilization in the setting of the preliminary injection of L-DOPA in a dose of 0.1 g per animal led to a more significant activation of the GGAKS in comparison with the effect of the stress influence without the injection of the preparation (Illustration 22). The amounts of the content of hydrocortisone in the blood were comparable only with the increase of the level of the hormone noted during a 10-hour immobilization. The maximum increase in the concentration of hydrocortisone was observed 6.5 hours after the injection of L-DOPA. The content of the hormone during this period amounted to 235% ($P<0.002$) relative to the initial level and 152% ($P<0.02$) relative to the amount of its concentration in the control experiment.

With the injection of L-DOPA in monkeys there was noted a clear tendency to a reduction in the testosterone content in the blood, which was noted throughout the 3 subsequent days after the injection of the preparation. The drop in the level of testosterone on the day of the experiment to a certain extent can be related to the influence of the experimental procedures, since in the control experiment with the placebo there was a similar dynamics in the change of the concentration of this hormone. However, 48 hours later in the control experiment there was observed a restoration of the level of testosterone to the initial amounts, while in monkeys with the injection of L-DOPA during this period there was noted a pronounced tendency for a reduction in the concentration of androgen in the blood.

During stress in a setting of L-DOPA injection there was noted a reduction in the concentration of testosterone in the blood similar to the control (stress in a setting of placebo). However in the following 2 days the content of testosterone remained low and amounted to 84 and 66% ($P<0.05$) relative to the initial values, while in the control group of monkeys 48 hours later there was noted a normalization of the concentration of the hormone in the blood.

Thus, the injection of L-DOPA leads to a pronounced activation of the GGAKS in monkeys under conditions of physiological calm and sharply increases the reaction of this system to the action of the stress irritant. The influence of L-DOPA on the endocrine function of the testes is manifest to a significantly lesser degree. In comparing the data which reflect the dynamics of the change of the content of gluco-corticoids in the blood and the excretion of KA in urine under the influence of L-DOPA in intact monkeys and under conditions of stress one can conclude that the increase of the secretion of gluco-corticoids observed is caused by an increase of KA production. The results of the research performed are a further confirmation of the participation of KA in the activation of the GGAKS under stress. The observed tendency toward a reduction in the secretory activity of the testes in monkeys with the injection of L-DOPA can be explained by a disturbance of the blood flow in these glands under the influence of the increase burst of KA.

One method widely used in clinical and experimental endocrinology, allowing us to reveal the functional reserves of the pituitary-adrenal complex, is a test using the injection of metopirone. Metopirone, which has an inhibiting effect on 11-beta-hydroxylase, suppressing the synthesis of the steroids in the adrenals at the stage of the transformation of 11-desoxyhydrocortisone into cortisone, leads to a reduction in the level of hydrocortisone in the blood and elicits an increase in the concentration of its predecessors--11-desoxyhydrocortisone, 17-hydroxylised derivatives, et al. (Komissarenko, Reznikov, 1972; Lange, et al., 1979; Schoeneshoefer, Fenner, 1981; Schoeneshoefer, et al., 1983; Sindler, et al., 1983; Goncharov, Butnev, 1984). Thus, the single injection of metopirone in a dose of 500 mg per os and 250 mg intravenously leads in sexually mature papio hamadryus males to a significant reduction in the content of hydrocortisone with a parallel increase in the level of 11-desoxyhydrocortisone, 17-oxypregnenolone, 17-oxyprogesterone and pregnenolone in the blood (Goncharov, Butnev, 1984; Tsulaya, 1985).

The preliminary introduction of metopirone 30 minutes before the beginning of a 2-hour immobilization of monkeys prevents the increase in the level of hydrocortisone in the blood elicited by stress (Tsulaya, 1985).

Taking into account the inhibiting influence of metopirone on the level of hydrocortisone in the blood, and the available data on the important role of gluco-corticoids in the production of KA under stress, we conducted research on the revelation of the functional activity of the SAS under severe ES in a setting of a preliminary injection of metopirone. With an intravenous injection of 250 mg of metopirone in monkeys there was noted an increase in the excretion of A in urine by 92% ($P<0.05$) relative to the initial values, a reduction in the discharge of NA, the level of which amounted to 36.1% ($P<0.02$) of the amount of its excretion before the injection of the preparation, and a statistically reliable reduction in the discharge of DA.

In experiments using immobilization in a setting of metopirone injection there was revealed a sharp increase in the secretion of A, the amount of which was 456% ($P<0.002$) relative to the initial level and which exceeded the corresponding values in the control experiment by 44% (Illustration 23). The excretion of NA in these experiments increased insignificantly and amounted to, on the average in 5 monkeys (53.94 9.32) n-moles/24 hours, which was 147% ($P<0.02$) higher than the initial values and 49% ($P<0.05$) less in comparison with the discharge of NA during the application of an analogous stress irritant without the injection of the preparation. Under stress in the setting of metopirone injection there is noted a certain reduction in the amount of the increase of DA. A reduction of the co-efficient NA/A from 1.71 (baseline) to 0.92 shows that the preliminary injection of metopirone leads, under conditions of stress, to a sharp predominance of the activation of the hormonal unit (A) over the mediator unit (NA).

The results obtained indicate that with the injection of metopirone along with a change in the balance of the steroid hormones there occurs a sufficiently pronounced change in the excretion of free KA in urine. One can assume that the observed changes in the functional activity of the SAS are the result either of a direct influence of metopirone on the activity of the KA-synthesizing and the KA-metabolizing enzymes, or are brought about by a reduction in the level of hydrocortisone in the blood, occurring under the influence of this preparation. To support the latter theory there is available data on the influence of cortico-steroids on the processes of the biosynthesis and metabolism of KA (Utevsky, Rasin, 1972; Shalyapina, Rakitskaya, 1976; Markey, et al.,

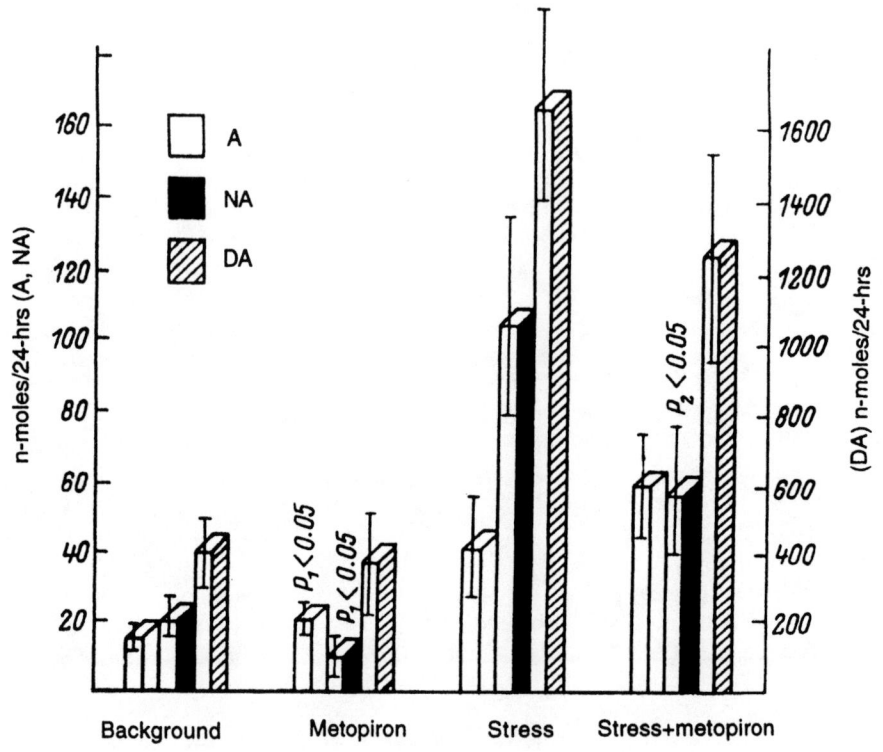

Illustration 23. Influence of metopirone on catecholamine excretion in monkeys during physiological calm and severe stress. Mathematical averages with reliable intervals.

Y-axis - , right - amounts of dofamin excretion (DA), n-moles/24-hrs'; left - amounts of adrenalin (A) and norepinephrine (NA) excretion, n-moles/24-hrs; P_2 - reliability of differences relative to stress.

1982; Haidarliu, et al., 1983). The sharp increase in the excretion of A observed in our experiments and the relative reduction in the reserve capabilities of the SAS in monkeys under conditions of stress in a setting of a low level of hydrocortisone in the blood agrees with available data indicating the increase in the exchange of KA and, in particular, the increase of the excretion of A and a decrease in the discharge of NA with a reduction in the level of cortico-sterone in rats with an injection of metopirone (Parvez, Parvez, 1972, 1978). The results of our research are also in accordance with the data of other authors, who revealed the rapid exhaustion of the catecholamine reserves both in the entire brain, and in the hypothalamus in rats under conditions of stress in a setting of hypercorticism, elicited by an adrenalectomy (Shalyapina, Rakitskaya, 1976). Consequently, the observed peculiarities in the change of KA excretion under stress in a setting of metopirone injection in monkeys are caused by a reduction in the production of hydrocortisone, which is a factor not only limiting the speed of KA metabolism because of its inhibiting influence on the activity of monoaminoxydase and catecholamine transfer (Parvez, Parvez, 1972, 1978). But also participating in the support of the increased secretion of KA under stress (Ciaranello, 1979).

It should be noted that the revealed peculiarities of the restructuring of the hormonal balance in monkeys during a metopirone test under conditions of physiological calm and stress, were evidently the reason for the sufficiently pronounced disturbances in the behavior of the monkeys, which were manifest in a general slowdown, a sharp reduction in the locomotive activity, a lethargy and apathetic attitude, the absence of a reaction of external stimulants, etc..

To analyze the functional interrelationships between the endocrine function of the testes and the activity of the SAS we studied the dynamics of the excretion of KA and VMK in monkeys with an injection of luliberin (Chirkov, et al., 1980) and chorionic gonadotropine under conditions of physiological calm and stress.

An intravenous injection of LG-RG led to a reduction in the amount of the excretion of KA and their metabolites in the daily urine, while an injection of a physiological solution increased the discharge of adrenaline ($P<0.02$), norepinephrine ($P<0.02$) and VMK, which reflected the development of the stress reaction to the experimental procedures, related to the clamping of the animal, the injection, and the intravenous injection (Illustration 24). With an injection of LG-RG the level of the discharge of A was reduced by 50.5%, of NA--by 26.3% and VMK--by 35.7%, in comparison with their initial values.

A similar but less pronounced decrease in the excretion of KA and VMK was noted with the injection of CKG (Illustration 24).

A preliminary introduction of luliberin and chorionic gonadotropine 30 minutes before the beginning of a 2 hour immobilization prevented to a significant degree the sharp increase in the level of KA and VMK (Illustration 24) elicited under these conditions. A more pronounced adrenolytic effect was also manifest with LG-RG. The level of the discharge of adrenaline, NA and VMK under stress in a setting of LG-RG was four times ($P<0.0001$), 4.4 times ($P<0.0001$) and 2.3 times ($P<0.0001$) less when compared with the corresponding amount in monkeys subjected to 2 hour immobilization without the injection of the

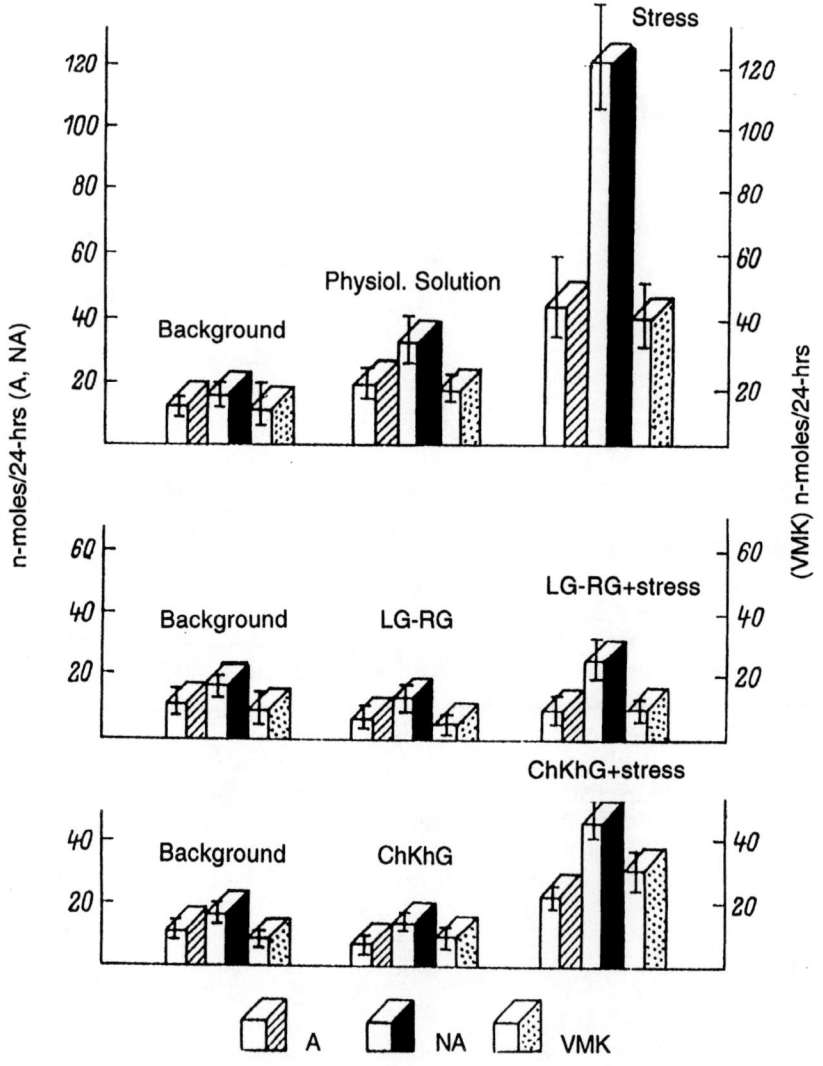

Illustration 24. Effect of luliberin (LG-RG) and chorionic gonadotropin (ChKhG) on the excretion of adrenalin (A), norepinephrine (NA) and vanilil-phenyl-glycolic acid (VMK) in monkeys under physiological calm and stress. Mathematical averages with reliable intervals.

Y-axis - , right - VMK excretion, m-moles/24-hrs; left - excretion of A and NA, n-moles/24-hrs.

preparation (the control). The preliminary injection of CKG also led to a smaller increase in the level of excretion of KA and VMK: the discharge of A, NA and VMK was 2.3 times ($P<0.001$), 3.2 times ($P<0.0001$) and 1.6 times ($P<0.0001$) less than the control, respectively.

The reduction in the activity of the sympatho-adrenal system under the influence of luliberin and chorionic gonadotropine, observed in our experiments, indicate the close interrelationship of the central and peripheral adrenergic processes with the functional state of the hypothalamus-pituitary-testes complex both under conditions of physiological calm and in the development of the stress reaction. A similar dynamic of the secretory activity of the SAS was revealed with the injection of luliberin in healthy males (Van Loon, 1978). The observed reduction in the level of KA circulating in the blood the authors explained by the possible influence of LG-RG on the presynaptic receptors of the catecholaminergic neurons.

The data obtained allow us to assume that along with the direct influence of luliberin on the functional state of the central and peripheral adrenergic structures its effect may be added to the influence of an increased level of androgens in the blood on the structures. This finds confirmation in the reduction of the level of KA excretion, observed in our experiments with the injection of CKG, eliciting a significant increase in the concentration of testosterone in the blood, as well as in results which points to a reduction in the content of KA in the tissues of animals with the injection of androgens (Bakhova, 1975).

As a result of the total research of the functional activity of the SAS and the dynamics of the content of gluco-corticoids in the blood in monkeys, conducted in various series of experiments, we can make a conclusion about the activating influence of catecholamines on the GGAKS under stress. This is indicated first of all by data obtained during the analysis of the interrelationships of the functional state of the sympatho-adrenal and the hypothalamus-pituitary-adreno-cortical systems under conditions of emotional strain, developing during the adaptation of monkeys to confinement in individual cages, and also under conditions of severe prolonged and repeated ES. With an increase in the intensity of the stress irritant or an increase in its exposition proportional to the increase of KA excretion there is noted an increase in the adreno-cortical activity. Secondly, with the injection of monkeys of a preparation of L-DOPA, increasing the production of KA in intact monkeys and increasing the activation of the SAS under stress, there is seen a pronounced statistically reliable increase in the level of hydrocortisone under conditions of physiological calm and a more significant increase in the adreno-cortical activity in response to immobilization in comparison with the reaction of the GGAKS under analogous stress applications without the injection of the preparation.

These results confirm the point of view of many authors about the activating influence of the central and peripheral units of the SAS on the secretion of gluco-corticoids under stress (Beitnis, et al., 1973; Shalyapina, 1976; Shalyapina, Rakitskaya, 1976; Johnstone, Ferrier, 1980).

Chapter 7

CORRECTION OF THE NEUROENDOCRINE IMPULSES UNDER EMOTIONAL STRESS IN MONKEYS WITH THE HELP OF PYRROXAN

In the arsenal of means which increase the resistance of the organism to the influence of stress factors, a special place belongs to psychotropic preparations as a result of their ability to selectively effect the fundamental psychopathological manifestations of emotional stress (Gubachev, et al., 1976; Stanishevskaya, Mezentseva, 1977; Valdman, et al., 1979).

In recent years thanks to the achievements of molecular and biochemical pharmacology there has appeared research allowing us to deeply penetrate the intimate mechanisms of the interaction of psychotropic preparations with the neurochemical substratus and thereby approach an understanding of the neurochemical bases for the influence of these preparations on the nature of psychic processes. A significant contribution to the development of the concepts of the mechanisms of the influence of psychotropic preparations is made by research exploring the influence of these preparations on the system of the synthesis and exchange of KA, the change of the neuronal transport and the seizure of adrenergic mediator, the number and affinity of adreno-receptors (Arefolov, 1982; Valdman, 1982).

Particular importance is assumed by research in which, along with the study of the role of the processes of adrenergic mediation in the realization of psychotropic effect, a place is given to the analysis of the functional state of other mediator and neuroendocrine systems, which have no less influence on the integral reaction of the organism (Arefolov, 1982; Valdman, 1982; Ignatov, et al., 1982; Kozlovskaya, et al., 1982; Poshivalov, 1982). Thus, the study of the role of the GAMK-ergic processes in the realization of the anti-stress influence of the preparations of a benzo-diazepin order manifest in the prevention of behavior and somatic disturbances, allowed Ignatov, et al. (1982) to establish the important role of GAMK-ergic mechanisms in the regulation of stress. In this the theory was promoted that one of the mechanisms of the neurochemical adaptation to stress is the suppression of the GAMK-transaminase, which prohibits the development of the change of the energetic metabolism of the brain and can be seen as one of the means of adapting to stress.

Recently the research of the psychotropic effects of neuropeptides is attracting enormous attention. It has been established that neuropeptides, simultaneously executing neuromediator, modulator and hormonal functions command a broad spectrum of biological activity and can have both a stimulating and a tranquilizing influence of the behavior of animals in extreme situations, manifesting a pronounced anti-stress effect (Vedernikova, Maisky, 1981; Kozlovskaya, et al., 1982; Poshivalov, 1982; Yumatov, 1983). The more clear modulating influence of the peptides is revealed in a setting of the disintegration of behavior, caused by the development of ES. In this it has been shown that the anti-stress effect discovered in certain peptides is closely tied to the change in the activity of the KA-ergic processes of the brain.

The results of the research of the effects of peptides and their combinations with psychotropic preparations have shown that the very effect of the influence of the neuropeptides on certain forms of behavior is temporary, and can increase or decrease under the influence of well-known psychotropic preparations (Poshivalov, 1982). However, the use of many psychotropic preparations, in particular tranquilizers and anti-depressants, for the purpose of preventing the development of ES reduces the intellectual activity of man, suppresses his psycho-emotional sphere and in a number of cases leads to the appearance of psychological and even physical dependence (Gubachev, Stabrovsky, 1981).

Due to this the search and use of new neurotrophic preparations which have a pronounced anti-stress influence but which do not reduce the intellectual activity of man and do not elicit the development of psychological dependence to their use, takes on enormous significance. These requirements are satisfied by an original national preparation of pyrroxan, which has primarily a central alpha-adrenal-blocking action. There is evidence of the positive effects of pyrroxan in therapeutic, psychiatric and neurological clinics: In vegeto-vasal dystonia and hypertensive crises, resulting from the pronounced predominance of the sympathetic tone, during alcohol abstention and emotional stress without a reduction of a level of psychic activity or intellectual activity of man (Martirosyàn, et al., 1976; Zonis, Brin, 1977; Ibatov, 1979).

The normalization of psychosomatic functions of the organism under the influence of pyrroxan can be explained by its adreno-blocking action, which prevents the increased injection of KA under stress and restores the content of biogenic amines in tissues during pathological states. This is confirmed by experimental data which shows that the injection of pyrroxan prevents excessive increase of the discharge of KA and normalizes the level of the hormones. Thus, a sharp increase in the excretion of KA, caused by a tetanus intoxication, is prevented by an injection of pyrroxan (Alexevich, Kun, 1981). Under the influence of this preparation in castrated rats the level of glycogen in the liver was increased and the mass of the adrenal glands was restored (Poskalenko, et al., 1980). In the studies of Aliev and Khassan (1977) the daily use of pyrroxan prevented the development of chronic stress in lactating rats, led to a reduction in the level of 11-oxs and noticeably increased the formation of prolactine and the growth hormone. The positive effect of pyrroxan on the functional state of the endocrine glands, observed in these studies, the authors relate to a change in the hypothalamic control. However, the question of the neuromediator specificity of the hypothalamic regulation (KRF) of the pituitary-adreno-cortical system at the present time has not completely been explained, while the available data from monographs, articles and individual works are very contradictory and point both to an increase in the secretion of KRF and AKTG during stimulation of the hypothalamic alpha-adreno receptors, and to the inhibition of the KRF-activity in response to the central release of A and NA (Ganong, et al., 1976; Sapronov, 1980; Robu, 1982).

At the present time there are studies which show the possibility of preventing the activation of the pituitary-adreno-cortical system under conditions of stress with the use of neurotropic preparations with various mechanism of influence. In addition, many neurotropic preparations, which slow down the activation of the GGAKS under stress, cannot be used as anti-stress preparations because of their pronounced various psychotropic effects, the manifestation of which is undesirable in a healthy man.

One of the hopeful directions for the prophylactics of the development of experimental dystrophic injuries to the heart and stomach, caused by the application of extreme irritants, is the use of neurotropic blockers, which prevent the excessive release of norepinephrine from the adrenergic terminals and which prevent the exhaustion of its supplies in these organs (Zavodskaya, et al., 1977; Zabrodin, Lebedev, 1979).

Due to this of special interest is the study of the influence of alpha-adrenal blockers pyrroxan on the functional activity of the SAS and the steroid producing glands in intact monkeys with a single injection this preparation, as well as under conditions of severe stress.

Experiments were conducted on intact sexually mature papio hamadryus males (Group 5, N=5), previously adapted to confinement in individual cages and to conditions of the experiment. Experiments were conducted with the intermuscular injection of pyrroxan in a dose of 1 mg/kg, as well as with the injection of pyrroxan in the same dose 30 minutes before the beginning of a 2 hour immobilization. Experiments with an injection of an isotonic mixture under conditions of physiological calm and stress were used as a control.

All experiments included an intramuscular injection of pyrroxan in a dose of 1 mg/kg in a 1% mixture. The chose of the dose was based on data we had obtained previously about the study of the influence of this preparation on the function of the cardio-vascular system and the hormonal balance under conditions of modelling the neurogenic arterial hypertension in papio hamadryus males (Chirkov, et al., 1980). In this the injection of the preparation in doses of 1.0 mg/kg and 1.5 mg/kg prevented the development of disturbances on the part of the cardiovascular system and had an equal sympatholytic effect, reducing

Table 10. Excretion of adrenalin, norepinephrine and vanilil-phenyl-glycolic acid in urine of male Papio hamadryas during a single injection of pyrroxan (Mm)

Injection	Background	Experiment
adrenalin (n-moles/24-hrs)		
physiological solution	8.52±0.6 $P_1<0.0001$	19.42±1.29 $P_1<0.05$
pyroxan, 1.0mg/kg	9.80±1.03	11.15±0.49 $P_2<0.001$
norepinephrine (n-moles/24-hrs)		
physiological solution	18.52±0.58	31.31±1.18 $P_1<0.0001$
pyroxan, 1.0mg/kg	17.15±1.43	16.76±0.74 $P_2<0.0001$
Vanilil-phenyl-glycolic acid (n-moles/24-hrs)		
physiological solution	6.06±0.20	8.08±0.88
pyroxan, 1.0mg/kg	6.73±0.23	5.97±0.19 $P_2<0.002$

Note. P_1 - reliability of differences relative to the background; P_2 - reliability of the differences relative to the physiological solution.

Table 11. Daily excretions of adrenalin, norepinephrine and vanilil-phenyl-glycolic acid in urine of monkeys under stress against a background of preliminary injection of pyrroxan.

Experimental conditions	Initial amounts	Experiment
	adrenalin (n-moles/24-hrs)	
pyroxan + stress	8.81±0.68	23.38±1.35 $P_1<0.0001$ $P_2<0.0001$
Stress (control)	9.63±0.84	48.38±2.68 $P_1<0.0001$
	norepinephrine (n-moles/24-hrs)	
pyroxan + stress	17.57±1.31	35.54±2.24 $P_1<0.001$ $P_2<0.0001$
Stress (control)	16.1±1.75	96.05±4.21 $P_1<0.0001$
	Vanilil-phenyl-glycolic acid (n-moles/24-hrs)	
pyroxan + stress	6.83±0.19	20.24±2.02 $P_1<0.001$ $P_2<0.0001$
Stress (control)	7.46±0.56	48.38±2.99 $P_1<0.0001$

Note. Here and in table 12, P_1 - reliability of differences relative to initial amounts; P_2 - reliability of the differences relative to the control. Pirroxan injected in a dose of 1.0 mg/kg.

in an equal degree the level of the response reaction of SAS under conditions of cardio-pathogenic emotional stress in the given species of monkey. At the same time the application of pyrroxan in a smaller dose (0.5 mg/kg) was insufficiently effective and did not prevent the pathological manifestation of stress.

The functional state of the SAS was evaluated according to the excretion of KA and VMK in the daily urine, a collection of which was made in each series of experiments throughout a period of 5 days (before the experiment, the day of the experiment, 3 days in the

Table 12. Content of hydrocortisone and testosterone in the blood of monkeys with an injection of pyrroxan under physiological calm and stress.

Experimental conditions	Time of blood draw			
	0	30 min.	2.5 hrs	6.5 hrs
hydrocortisone (n-moles/ltr)				
Injection of physiological solution	825±112	906±153	663±99	257±93 $P_1<0.01$
Injection of pyroxan	844±108	914±113	778±103	237±80 $P_1<0.002$
Pyroxan + stress	766±56	860±52	1299±153 $P_1<0.02$ $P_2<0.05$	1191±185
Stress (control)	897±113	935±137	1961±184 $P_1<0.02$	1238±119
Testosterone (n-moles/ltr)				
Injection of physiological solution	27.91±4.21	25.32±5.8	14.53±3.8 $P_1<0.05$	21.63±6.18
Injection of pyroxan	26.71±5.98	23.91±6.9	18.96±3.3 $P_1<0.002$	26.96±8.16
Pyroxan + stress	29.38±5.48	21.37±3.1	16.23±2.2 $P_2<0.02$	23.27±4.72
Stress (control)	25.89±3.21	22.50±4.2	13.31±2.55 $P_1<0.02$	7.68±2.31 $P<0.002$

restoration period). The steroid hormones were determined in the plasma of peripheral blood, which was collected according to the following schedule: before the injection, at the following intervals after the injection of the preparation—30 minutes, 2.5, 6.5, 24, 48 and 72 hours.

The single injection of pyrroxan (1 mg/kg) in intact monkeys did not change the daily excretion of KA and only insignificantly reduced the level of VMK when compared with the amounts of their discharge in the preceding days, while the injection of a physiological mixture (the control) was accompanied by a reliable increase in the daily excretion of A and NA ($P<0.0001$) and by an insignificant increase in the discharge of VMK relative to its initial values (Table 10). The activation of the SAS observed in the control experiment reflects the insignificant (in the degree of expression) psycho-emotional reaction to the experimental procedures, the development of which is not noted with the injection of pyrroxan.

The preliminary 1 time injection of pyrroxan to a significant degree prevented the increase in the level of KA and VMK in response to the 2 hour immobilization when compared with the amounts of their discharge during a similar stress application without the injection of the preparation (Table 11). Under immobilization with pyrroxan the excretion of A, NA and VMK amounted to 50%, 37% and 40% of their respective amounts in the control experiment.

The analysis of the functional state of the steroid-producing glands, conducted in our experiments, showed that the injection of pyrroxan does not have an influence on the level of hydrocortisone in the blood, a reduction in the concentration of which during the evening corresponds to the daily rhythm of the content of this hormone in this species of monkey (Taranov, 1981). With the injection of pyrroxan in intact animals there was noted an insignificant and short-lived reduction in the concentration of testosterone in the blood, while in the control experiment with the injection of a physiological mixture this reduction was more pronounced and prolonged. The data obtained show that the injection of pyrroxan does not completely remove the inhibiting effect of the psycho-emotional stress, caused by experimental procedures, on the endocrine function of the testes in monkeys.

The increase in the concentration of hydrocortisone in response to a 2-hour immobilization, conducted 30 minutes after the injection of pyrroxan, was 34% ($P<0.05$) less in comparison with the level of hormone under similar stress application without the injection of the preparation. However, the increase in hydrocortisone in experiments using immobilization and pyrroxan was reliable ($P<0.02$) relative to the initial level. The protecting effect of pyrroxan was manifest also in preventing the inhibiting influence of stress on the secretary function of the testes. The content of testosterone in the blood 2 hours after the beginning of the application, which was conducted using the injection of the preparation, amounted to (16.23±2.21) n-moles/l, which was 21.9% higher in comparison with the control. 6.5 hours after the injection of the preparation under these conditions there was already noted a complete restoration of the level of androgen in the blood, while in the control experiment the reduction of the testosterone concentration during this period was maximal, and its level amounted to only 29.7% of the initial level ($P<0.002$).

Thus, a preliminary injection of pyrroxan prevents activation of the SAS, slows down the development of the pronounced adreno-cortical response and inhibits the deep suppression of the endocrine function of the testes during the action of the powerful psycho-emotional irritant, which a 2-hour immobilization is for monkeys. The observed effect was evidently caused primarily by pyrroxan blocking the central alpha-adrenergic structures, the excitement of which is one of the leading mechanisms in the activation of the hypothalamus-pituitary-adreno-cortical system (Robu, 1982).

The results of the research agree with available data which points to the normalization of the disturbances of the neuro-endocrine balance under the influence of pyrroxan in other laboratory animals in experimental stress situations (Veshchilova, 1975; Aliev, Khassan, 1977; Poskalenko, et al., 1980; Aleksevich, Kun, 1981). In addition the data obtained point to a significant increase in the level of hydrocortisone under conditions of a 2-hour immobilization with preliminary injection of pyrroxan, and they allow us to assume that activation of the pituitary-adrenal complex under conditions of stress is not the result of a selective excitement of the central adrenergic structures of the brain. This is in complete agreement with available data on the multi-mediator mechanism of the regulation of the pituitary-adrenal complex under conditions of stress (Ganong, et al., 1976; Sapronov, 1980).

The study of the state of the psycho-emotional sphere in monkeys under different stress situations with preliminary injection of pyrroxan, which we conducted with the help of the ethnological approach revealed its clear anti-stress influence according to the indicators of individual and communicative behavior, as well as the clearly pronounced effect during an increased intra-species aggression.

The possibility of the prevention of severe neuro-endocrine impulses and the normalization of behavior of monkeys under conditions of stress with the aid of a preliminary injection of pyrroxan, revealed in our experiments, points to the presence of a pronounced stress-protective effect of this preparation, which allows us to, taking into account the available data on the normalization of the psycho-emotional sphere under the influence of pyrroxan in man, recommend that this preparation be more widely used for the prophylactic purpose in the development of emotional stress in man and animal under the influence of various stress and extreme factors.

Conclusion

The large amount of experimental and clinical material available in scientific literature attests to the fact that the development of resistance to the influence of extreme irritants is accompanied by high energy expenditures and the stress of the regulatory systems of the organism. An exceptionally important role in the mobilization of energetic and structural resources belongs to the activity of the sympatho-adrenal and hypothalamus-pituitary-adreno-cortical systems--the leading neuroendocrine units, which provide the formation of the general non-specific reaction of the organism in response to the influence of the widest variety of stress and extreme factors of the environment.

The activation of these systems within the boundaries of homeostatic (physiological) regulation of the function prevents the development of specific pathological processes, thereby exerting a protective effect against stress injuries. The participation of catecholamines and gluco-corticoids in the immediate trigger reaction of the organism, in the regulation and coordination of the physiological processes, is realized thanks to the reactive changes of the functional state of the sympatho-adrenal and hypothalamus-pituitary-adreno-cortical system, which are expressed in an increase in the secretion of hormones and neuro-mediators, the appearance of their massive amounts into the blood and into other liquid media of the organism, with various accumulations in the various organs and tissues.

It has been shown that the regulation of the processes of adaptation on the part of the neuro-endocrine system is ensured not only by the effects of catecholamines and gluco-corticoids and by their interaction on a cellular level, but also depends on the influence of other neuro-mediators, hormones and their biologically active predecessors and metabolites, as well as neuro-peptides (Vedernikova, Maisky, 1981; Kosta, Trabukki, 1981; Meerson, 1981, 1984; Robu, 1982; Yumatov, 1983; Anokhina, 1984; Burov, Vedernikova, 1984; Chirkov, 1984; Tsulaya, 1985; Axelrod, Reisine, 1984). The continuity and harmony of the activation of the leading stress-realizing neuro-hormone system in their unbroken interaction with other neuro-endocrine and neuro-chemical systems, engaging in the dynamics of the stress reaction of the organism, is an important unit which determines the formation of the adaptive processes, and depends on many factors: the specifics and intensity of the stress irritants, the duration and frequency of their application, the length of intervals between individual applications, and also the initial state and phase of the daily activity of the organism.

The development of stress injuries under conditions of ES is seen on the whole as the result of excessive increase of the adaptive effect of stress, the transformation of its adaptive nature into an injurious one (Meerson, 1981, 1984; Rotenberg, Arshavsky, 1984). What's more, as leading factors in the transformation of the adaptive mechanism into a general unit of pathogenic stress injury, manifest in the development of various psychic and somato-visceral disturbances, there can appear various impulses of the function of the neurochemical systems of the brain, pronounced restructuring of the neuro-hormonal balance of the organism and changes in the nature of the interaction of neuro-mediators and hormones with the target organs on a cellular and molecular level.

It is necessary to also note that the resistance of the organisms to the action of psychogenic irritants depends on the genetically produced characteristics of the exchange of biogenic amines and the nature of their functional interaction with neuropeptides. Disturbances of the physiological equilibrium of the activity of various neurochemical structures of the brain and of the coordinated activity of the glands of internal secretion, caused by excessive increase of the stress application, are expressed in an excessive production and change in the amounts of the ratio of individual mediators and hormones, in the lengthy decrease of the level of secretion and in the complete drop-off of the physiological effect of a number of biologically active endogenic compounds, in the accumulation of the products of their metabolism and other manifestations of a neuro-hormonal imbalance. Thus, the excess of catecholamines along with the increase of their physiological effects on a cellular level under the influence of significantly increasing content of cortico-steroids in the tissues, which arises under conditions of severe or chronic emotional stress, is considered one of the real pathogenic mechanism of the development of various cardio-vascular illnesses in man.

Experimental studies in particular have shown that the formation of arteriosclerosis and resistant arterial hypertension of an emotiogenic origin requires the joint increase in the secretion of hormone of the cortical and brain matter of the adrenals, and furthermore the increase of the adrenergic effect in myocardia on a cellular and molecular level is accompanied by the development of a shortage of phosphorous compounds, rich in energy, by the reduction in the content of glycogen, and the effectiveness of the use of oxygen by mitochondria, by an accumulation of fatty acids, a disturbance of the membrane, a significant excess of Ca+2, by contractual injuries of the myofibril and the death of cells (Sudakov, 1976, 1981; Khomulo, 1982; Meerson, 1984).

On the other hand, the exhaustion of the tissue reserves of the adrenergic mediator (NA), which occurs after extreme strain of the sympathic nervous system, with the subsequent action of neurogenic irritants there occurs as an additional unit of pathogenic mechanisms of structural disturbances--neurogenic dystrophy of various organs (Zavodskaya, et al., 1977; Zabrodin, 1982). Taking into account the close interaction of catecholamines with other neuro-mediators and hormones in the achievement of the adaptive effect, the complex nature of the functional interrelations of the sympatho-adrenal system with the activity of the hypothalamus-pituitary-adreno-cortical, gonad, thyroid systems and other neuro-endocrine complexes, as well as the biologically expedient, temporal irregularity of the manifestation of adaptive phase of restructuring of their reaction, one can assume that in the formation of one or another psycho-pathological process, and in the disturbance of the functions of the internal organs under conditions of emotional stress, a great significance belongs to the steroid and protein hormones, to the prostaglands, neuro-peptides and other biologically active substances, which change the adrenergic mediation.

Based on this the intensive study of the nature of the functional state of the leading units of the neuro-endocrine system under stress is not the sole purpose of the research conducted by us and by other authors on a molecular level but serve as the chief task of experimental medicine, consisting of the revelation of the mechanism of the adaptive reactions of the entire organism, and answers one of the more important provisions of clinical medicine on the necessity of a more economic use of the organism's own defense reactions for the increase of its resistance to unfavorable factors of the environment, for the purpose of prophylactics and treatment of illnesses. The material presented in this book is dedicated to the analysis of the functional state of the main regulatory systems--the SAS and GGAKS under stress, as well as to the study of that important unit of induction of anabolic processes, the endocrine function of the testes. The clarification of the dynamics of the hormonal activity of the gonads was of special interest for the differentiation of the phase of physiological stress of the regulatory system under a weak and measured intensity of the stress application from overstress, which is the beginning of the formation of the characteristic pathological manifestations of stress in response to the action of a powerful stress irritant, in as much as it is well-known that the adaptive forms are only those forms of active purposeful reaction of the organism, which secure the induction of excess anabolic processes (Rotenberg, Arshavsky, 1984). The provision on the conjugate action of catecholamines and steroids on the physiological processes, and also the difference of the spatial-temporal parameters and energetic supply of stress and adaptation make necessary the further development of adequate criteria of these states, which, on the one hand, should open the way for prognostic evaluation of the overall dynamics of the reaction toward an increase in the resistance or pathology, and on the other hand, will allow us to select tactics for increasing the adaptive ability of the organism.

The significant amount of facts about the great morpho-functional similarity between the physiological processes in monkeys and in man, which has accumulated in the literature, the great and clear emotional expressiveness of the behavior reactions of these animals, along with the similarity of the direction and the dynamics of the adaptive restructuring of the functions of the sympatho-adrenal and hypothalamus-pituitary-adreno-cortical, gonad systems in papio hamadryus males under emotional stress, and the hormonal manifestations of the stress reactions in man, revealed in our experiments, allows us to come to the conclusion of the expediency of the use of monkeys in psycho-endocrinological research on the study of the mechanisms of stress and adaptation, since the data obtained in these experiments, with a great degree of approximation, reflect the state of the vegeto-somatic function in the organism of healthy people, found under stress situations.

In our research on the study of the peculiarities of the pathogenesis of emotional stress in papio hamadryus males, particular attention was given to the nature of the inner- and intra-system ties and restructurings of the functioning of the sympatho-adrenal system and the steroid producing glands. In summarizing the data obtained it should be emphasized that an evaluation of the general direction of the development of stress reactions and the formation of the processes of adaptation, just as the degree of expression of the pathological phenomena, requires a complex approach, including both the study of functional state of sympatho-adrenal and hypothalamus-pituitary-adreno-cortical, gonad systems, and the simultaneous study of the somato-visceral function and the conduction of ethological analysis of individual and communicative behavior of monkeys. This allows a comparison of the change of the neuro-endocrine balance and the nature of the neuro-humoral-hormonal interrelationships with the dynamics of the formation of pathological syndromes of stress, their expression and resistance.

The use of immobilization as a stress irritant, which we used earlier for modeling of the neurogenic pathology in this species of monkey, gives us the opportunity to reveal the most common regularities of the neuro-hormonal shifts in the mechanisms of the development of the system neurotic somato-visceral pathology. It is important to note that the established dependence of the degree of expression of the reaction of the sympatho-adrenal system, the adrenal cortex and the testes to the influence of stress irritants on the phase of daily activity of the organism, reflects a close interrelationship between the phenomenon of the periodicity of the physiological functions and stress, which attests not only to the manifestation of the common biological dependency of the reaction on the initial functional state of the organism, but also confirms on the whole the idea of the selective injury of the functional systems under emotional stress, according to which that system which is at the moment of its application in the activated state, is considered the most vulnerable to the influence of the stress (Startsev, 1976).

It has been established that the development of emotional stress in morning hours, which correspond to the maximum overall motion activity of the monkeys and to the greatest secretory activity of the adrenal cortex, is accompanied by a less pronounced reactive change of the state of the leading neuroendocrine complexes and in addition by a more significant and prolonged disturbance of the individual and group behavior, in comparison with the response reaction to immobilization in the evening hours, which was characterized by a high level neurohormonal activity and by minimal shifts in the behavior of the animals. This indicates that securing the adaptation to the influence of the powerful stress irritants requires not only a high level of glucocorticoids in the blood, but a sufficient degree of reactivity of the response of the GGAKS, which in the acrophase of daily secretion of these hormones was reduced. An analysis of the hormonal activity of the testes showed that in the post-period of morning stress there is noted a more lengthy suppression of the secretion of androgen, while in response to the influence of stress in the evening hours the suppression of the endocrine function of the gonads was transitory, and normalized on the following day, and correlated to insignificant and brief shifts in the behavior of the monkeys.

The results of our research indicate that the appearance of psychopathological states requires not only a change in the concentration and disturbance in the normal ratio of hormones and mediators, but a certain level of initial functional activity of the organism, caused in its turn by the synchronization of the C-rhythms of the physiological processes and by the regulating influence of the daily periodical secretion of hormones with the geophysical units of time.

The reduction of the psychic and motion activity in the post-period of emotional stress, produced during morning hours, we characterized as a depressive-type state, in the mechanisms of which, along with an exhaustion of the central and peripheral catecholaminergic depot as a result of a significant injection of catecholamine and along with a change in the ratios between them and the influence of the excessive secretion of gluco-corticoids on the processes of adrenergic mediation a great role, in our opinion, belongs to the sharp suppression of the endocrine function of the testes and to the drop in the level of testosterone in the blood, which is indicated in part by the presence of a direct correlation in the dynamics of the restoration of its concentration with the normalization of behavior.

Data from the conducted experiments allows us to conclude that the suppression of androgenopoesis in the testes is in some instances a more sensitive indicator of the development of stress reactions and according to the length of the response, as was shown in experiments using 10-hour and repeated immobilizations, it exceeds the response reactions of the sympath-adrenal and hypothalamus-pituitary-adreno-cortical systems. In spite of the adaptive nature of this reaction, which evidently reflects the absence of the need of the tissues for androgens in the initial phase of the stress reaction, the excessive and prolonged suppression of the function of the male sex glands, leading to a cessation of the anabolic effect of testosterone, is an extremely negative factor, which limits the possibility for the formation of anabolic processes and the adaptation of the organism to the influence of extreme irritants. The deep and prolonged suppression of androgens, revealed in experiments using 10-hour immobilization and daily 2-hour immobilizations for periods of 6 days, is comparable only with the response reaction to grave and massive surgical invasions in men.

The established phenomenon of psychogenic, reversible "castration" in sexually mature monkeys indicates the important role of emotional overstrain in the disturbance of the hormonal function of the testes and in the reduction of the level of anabolic hormones in the blood. The revealed regularities of the restructuring of the hormonal balance in experiments on monkeys and, in particular, the prolonged reduction of testosterone concentration in the blood in response to the influence of psychogenic stimulants, aids in the deeper understanding of the role of neuroendocrine shifts in the pathogenesis of various neurotic, psychotic and psychosomatic disturbances in man. Proof of this are the data we obtained in experiments on sexually immature papio hamadryus males, on the development of a resistant form of neurogenic arterial hypertension under conditions of repeated emotional stress. A more significant difference in the initial state of the hormonal balance in sexually immature monkeys, which presented a unique biological model of the physiological hypofunction of the testes, was the borderline low testosterone content in the blood when compared with adult animals, in whom in the response to analogous stress applications in a setting of high initial hormonal activity of the gonads there was noted only a transitory increase in the amounts of the arterial pressure (Illustration 25). A comparative analysis of the interhormonal interrelationships showed that in spite of the similar and adaptive nature of the

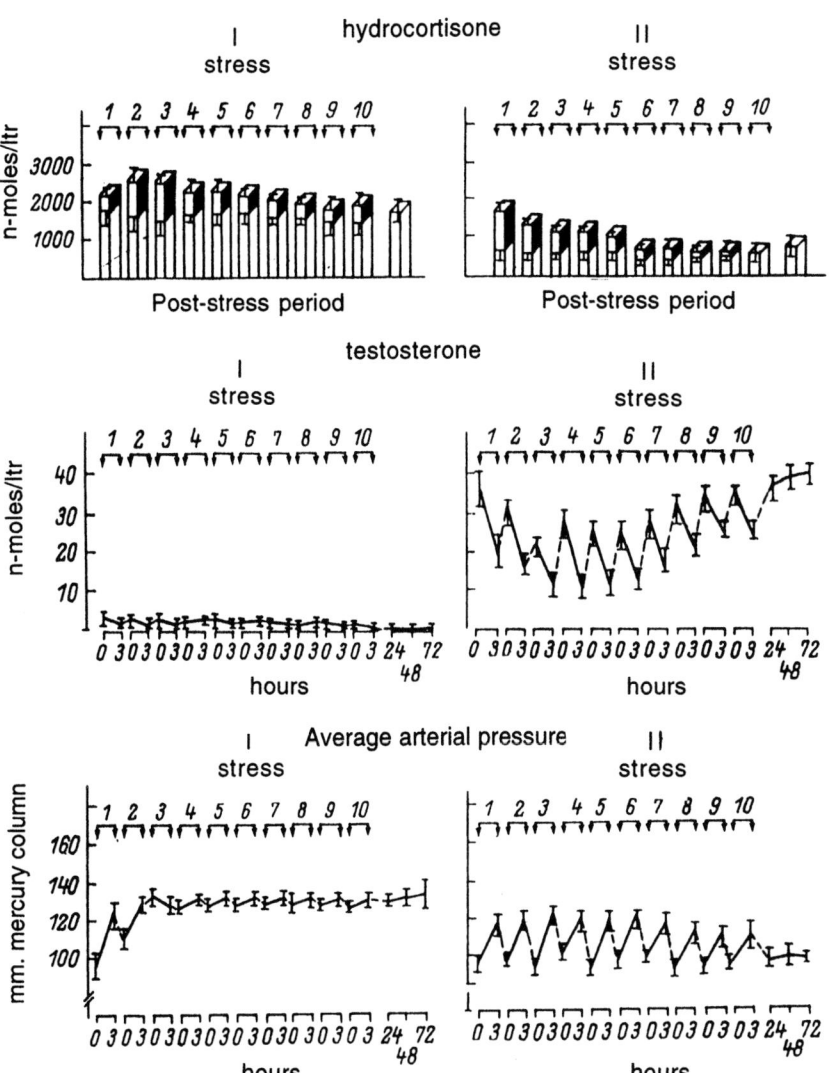

Illustration 25. Dynamics of change in content of hydrocortisone, testosterone in the blood and of the average arterial pressure (AD) in sexually immature monkeys under repeated emotional stress.

Y-axis - - content of steroid hormones, n-moles/ltr, AD, mm. mercury column.
X-axis - 0 - before immobilization, 3 - three hours after start of immobilization; I - sexually mature monkeys, II - sexually immature monkeys ; clear columns - content of hydrocortisone before immobilization; black columns - content of hydrocortisone three hours after start of immobilization.

processes of restructuring the functioning of the sympatho-adrenal and hypothalamus-pituitary-adreno-cortical systems, observed in adults and sexually immature individuals under conditions of repeated stress, in the later throughout the entire period of stress applications there were noted significantly greater amounts of dofamine and signs of a reduction in the reserves of the SAS. Based on the results of this research and on available data on the inhibiting influence of androgens on the exchange of catecholamines, we proposed the theory that a low activity of the endocrine function of the testes is one of the factors causing the development of neurogenic arterial hypertension in sexually immature organism. This theory is supported by data of other authors, pointing to a certain mutual relationship between the reduction in androgen production in men and cardiovascular illnesses (Gerasimova, et al., 1978; Gubachev, Stabrovsky, 1981). Considering the close interaction of catecholamines and testosterone with neuropeptides and prostaglandins, as well as the pronounced anti-stress and hypotensive effect of the latter, one can assume that in the mechanisms of the increase of the pressor adrenergic influences of stress on the circulation system a certain role belongs to the shortcomings of the sub-strata or enzymatic supply of the exchange of neuropeptides and protaglandins-- the natural anti-stress systems of the organism.

In summarizing the results of the research we conducted one can conclude that the complex study of the functioning of the sympatho-adrenal and pituitary-adreno-cortical gonad systems with the simultaneous analysis of the inner-system and inter-hormonal relationships helps us obtain hopeful information on the state of the psycho-physiological reactivity of animals in a sufficiently broad range: from the insignificant degree of emotional strain in response to changes of the customary conditions of confinement and the influence of various situational factors, to the clearly pronounced emotional stress, which forms under the influence of more intensive stress irritants. The simultaneous determination of catecholamine, gluco-corticoid and androgen content allowed us to reveal the dependence of the nature and degree of the expression of the reaction of the sympatho-adrenal system and the steroid producing glands on the temporal structure of the application of stress irritants, which includes their duration and frequency of action, as well as the length of intervals between them.

The established dependence reflects on the whole the correspondence of the stress activation of the SAS to the nature and expression of the hormonal reactions of the steroid producing glands. Thus, the excessive activation of the SAS, observed according to the strengthening or increase in the duration of the action of the stress stimulai, assists the greater excitement of the pituitary-adreno-cortical system and a deep, prolonged suppression of the endocrine function of the testes. Under conditions of repeated ES a maximum expression of the restructuring of the neurohormonal balance in monkeys is seen with a daily 2-hour immobilization, while the imposition of preliminary stress application or the increase of intervals between immobilizations leads to a relative reduction in the neurohormonal reaction. However during the period after the end of the stress applications or during the stress with injection of hormonal and neurotropic preparations, the relationship of the amounts of KA injection to the content of steroid hormones can be broken, reflecting on the one hand distinguishing characteristics of the dynamics of the adaptive processes of the functioning of the SAS, GGAKS and GGGS, and on the other hand--the complex nature of their functional interaction in the realization of the stress reaction and the processes of restoration.

As the result of experiments conducted it was established that under the influence of daily 2-hour immobilizations in sexually mature papio hamadryus males there occurs an excessive activation of the sympatho-adrenal system, a sharp increase in the concentration of hydrocortisone and its predecessors, reaching extremely high values, with a simultaneous deep and lengthy drop in the content of testosterone in the blood. The relative normalization of the level of gluco-corticoids, which occurs after the end of each stress application, furthermore does not coincide with the restoration of testosterone concentration, the amounts of which remained extremely low even throughout the 2 following days after the final immobilization. It should be noted that in the dynamics of repeated ES there were noted restructurings, adaptive in their nature, of the functioning of the sympatho-adrenal and hypothalamus-pituitary-adreno-cortical systems, which was manifest in the relative reduction of the stress amounts of catecholamine excretion, with the increase of the discharge of vanilil-phenyl-glycolic acid, the decrease in the time required for a maximum increase and the restoration of the gluco-corticoid content of the blood in response to repeated immobilizations, the development of positive correlations between hydrocortisone and its precursors and a number of other signs pointing to the engagement of the natural breaking systems, limiting the excessive excitement of the SAS and GGAKS.

In spite of the adaptive direction of the change of the neurohormonal balance on the whole and the response reactions of the sympatho-adrenal and pituitary-adreno-cortical systems, it is obvious that the achievement of resistance of the organism under these conditions is realized by means of excessive tension of the neuroendocrine mechanisms of regulation and great energy expenditures.

As shown by the research of Garkavi, et al. (1977), there is the possibility of an increase in non-specific resistance of the organism by means of the application of weak or measured stimulants without the element of harm and neurohumoral disturbances, characteristic for

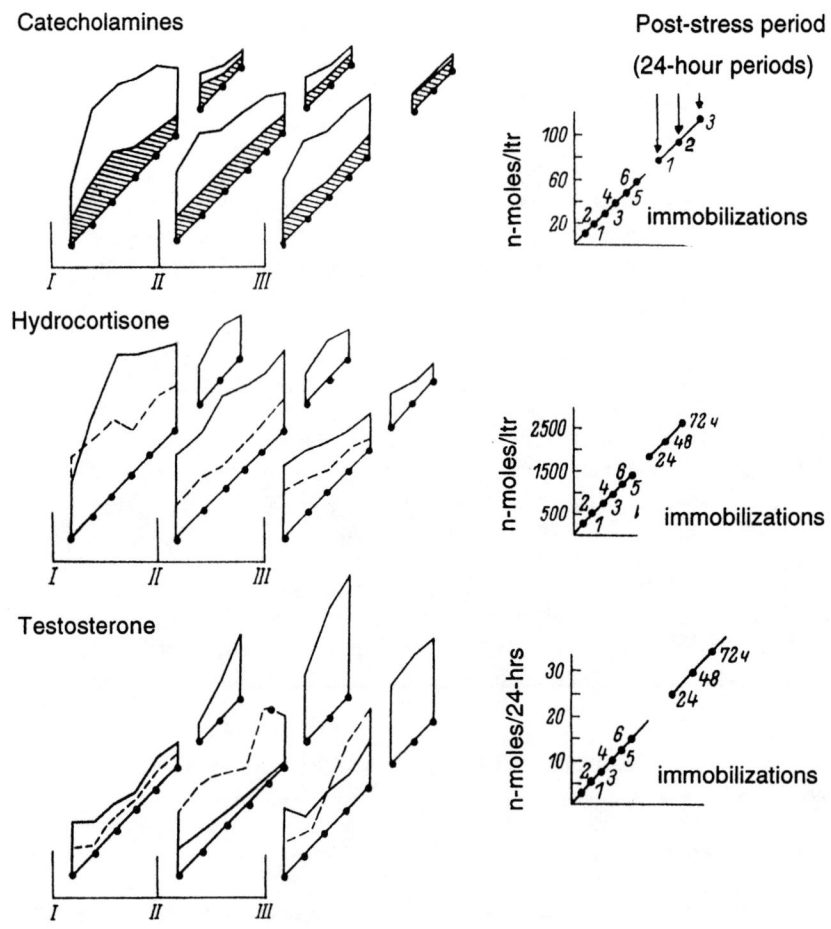

Illustration 26. Nature of adaptive restructurings of the neurohormonal balance under repeated emotional stress depending on the regimen of application of the stress irritants.

Y-axis - amounts of excretion of adrenalin and norepinephrine, n-moles/24-hrs; concentration of hydrocortisone and testerosterone, n-moles/ltr; X-axis - : I - cycle of daily 2-hour immobilizations; II - cycle of daily 2-hour immobilizations with preliminary immobilization (preliminary immobilization not shown); III - cycle of immobilizations with intervals of 2, 24-hour periods; clear section of the plane - amounts of norepinephrine excretion; striped sections of the planes - amounts of adrenalin excretion; upper solid line of the plane - content of hydrocortisone and testosterone 2 hours after the start of immobilization; broken line - hormone content 6 hours after the start of immobilization.

stress. Of exceptional importance to us in theory and in practice was clarifying the possibility of a directed increase in the resistance of the regulator neuroendocrine systems themselves to the action of extreme irritants (immobilization), without changing their intensity or duration, but regulating the length of the intervals between repeated stress applications. As our research showed, this approach is very hopeful both in the plan of achieving a relatively rapid depression of response reactions of the sympatho adrenal and pituitary-adrenocortical systems, and in the increase of the speed of the development of the adaptive changes of their functioning and in the restoration of the normal level of catecholamines and steroid hormones during the post-periods of repeated stress irritants.

The application of preliminary stress irritant 3 days before the cycle of daily immobilizations leads to a significant decrease in the degree of activation of the sympatho-adrenal and hypothalamus-pituitary-adreno-cortical systems, and, what is particularly important, to a rapid and complete restoration of the initial concentration of testosterone in the blood 6 hours after the beginning of each repeat stress application. A great expression of the adaptive restructurings of the functioning of the SAS and GGAKS, even to a complete suppression of the adreno-cortical reaction under the transitory nature of the suppression of the hormonal activity of the gonads is noted with an increase in the duration of the intervals between immobilizations to 72 hours (Illustration 26). The established dependence of the degree of expression and the nature of the neuroendocrine response on the temporal schedule of the sequence of repeated stress stimulai attests to the possibility of a directed increase in the resistance of the regulatory systems themselves to emotiogenic influences, and can be seen as an experimental basis for the development of more economic training regimens of psychoemotional loads in man for the purpose of preventing overstress of the neuroendocrine system or the development of stress injuries.

Preventing the development of psychopathological manifestations of emotional stress can be achieved also without preliminary adaptation to the influence of the stress irritants, by way of a directed pharmacological correction of the disturbances of the activity of individual units of the regulatory systems with the aid of natural and synthetic chemical compounds, which selectively strengthen or block the activity of various neurochemical systems of the brain and of the glands of internal secretion. It should be emphasized that an analysis of the influence of psychopharmacological preparations on the complex structures of free inner-species behavior of animals is particularly important for the evaluation of their anti-stress action. The pharmaco-ethological approach, the object of study of which are various aspects of the zoo-social behavior and the brain neurochemical mechanisms

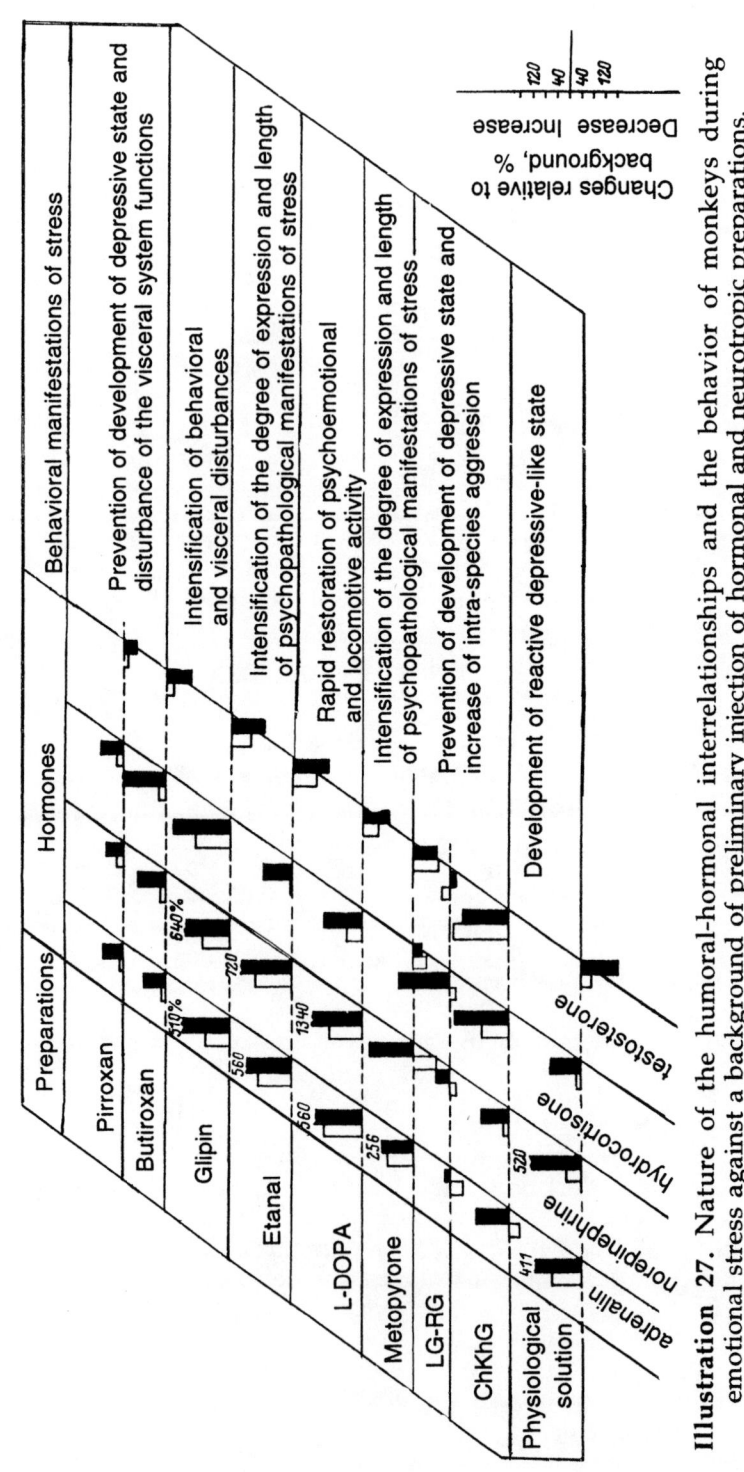

Illustration 27. Nature of the humoral-hormonal interrelationships and the behavior of monkeys during emotional stress against a background of preliminary injection of hormonal and neurotropic preparations.

Clear columns - changes in the excretion of catecholamines and concentration of steroidal hormones in the blood during injection of the preparations (control - injection of a physiological preparation); dark columns - change in the excretion of catecholamines and concentration of steroidal hormones in the blood under stress (2-hour immobilization) against a background of preliminary injection of the preparations (control - stress against a background of preliminary injection of a physiological solution).

which secure its integration, significantly expands the possibilities for experimental psychopharmacology and allows us to objectively evaluate the abilities of psychotropic means for the correction of abnormal forms of inner-species behavior of animals, arising during emotional stress (Valdman, et al., 1979; Valdman, Poshivalov, 1984). Using the ethnological method of analysis of individual and inner-species behavior of monkeys, along with the study of the function of the neuro-hormonal systems, has shown the promise of this complex trend both for revealing the mechanisms of the development of psychopathological manifestations of emotional stress and for evaluating the effects of anti-stress influence of the various psychopharmalogical means (Chirkov, et al., 1982, 1986b; Chirkov, Katsia, 1983; Butovskaya, et al., 1983; Chirkova, et al., 1983, 1986; Deryagina, et al., 1984; Lapin, et al., 1984; Startsev, et al., 1986). Illustration 27 presents a summary view of data on the influence of certain hormonal and psychotropic preparations on the functional state of the neurohormonal systems, evaluated together with the dynamics of the ethnological indicator of free group interaction under conditions of emotional stress.

From the broad arsenal of adaptagens, used to prevent ES, especially promising are the psychotropic preparations and biologically active substances, including neuropeptides, which, preventing the development of harmful consequences of the stress situations, do not suppress the intellectual activity of man and do not elicit the appearance of dependence on their use. The study of stress-protecting characteristics of our national preparation, the central alpha-adreno blocker pyrroxan in an experiment on monkeys allowed us to reveal the high effectiveness of the use of this preparation for purposes of preventing the development of depressive-type states, disturbances of the conditioned reflex activity, as well as a correction of the neuro-hormonal and visceral impulses under the influence of stress irritants, which allows us to recommend its broader application as an anti-stress preparation.

Based on these facts and available data we can conclude that one of the promising approaches to determine the development of medical-biological research on the problem of preventing pathological manifestations of emotional stress, is the search for physiological and pharmacological methods and means which allow us to realize a directed regulation of the processes of adaptation by way of a selective or complex application on the function of the leading neuro-hormonal systems of the organism, taking into account the nature of their interaction. Such an approach will require great efforts on the part of physiologists, neuroendocrinologists, psychopharmacologists, and other specialists enroute to a further in-depth study of the role of hormones and neuromediators, including substances of a peptide nature, in the regulation of psychic processes, emotions and behavior, the functions of various physiological systems and the defense forces of the organism.

BIBLIOGRAPHY

Airapetyants, M.G., Vein, A.M., *Neuroses in Experiments and in Clinic.* Moscow, 1982. 272 pp.

Aleksevich, Y.I., Kun, N.D., The Influence of Sympatholics of Pirroxan on the Excretion of Catecholamines during Experimental Tetanus Intoxication. *Pharmacology and Toxicology.* 1981. v. 66, no. 1. pp. 115-117.

Aleshin, B.V., Bondarenko, L.A., On the Mechanism of the Disturbance of Androgenopoesis under Stress. *Bulletin of Experimental Biology and Medicine,* 1982. v. 94, no. 7. pp. 98-100.

Aleshin, B.V., The Duality of Neurosecretary Mechanisms of the Hypothalamus and its Significance in the Regulation of the Endocrine Functions. *Achievements of Physiological Sciences* 1979. v. 10, no. 1. pp. 7-27.

Aliev, M.G., Khassan, G.A., The Action of Pirroxan on the Formation of the Pituitary Prolactin, the Growth Hormone and the Level of 11-OKS in the Blood in Lactating Rats, the Standard and under Conditions of Stress. *Materials of the Second Congress of Endocrinologists of the Ukrainian SSR.* Kiev, 1977. pp. 103-104.

Allikmets, L.K., Mutual Influence of L-Tryptophane and L-DOPA on the Behavior and Exchange of Mono-Amines in the Brain of Rats. *Academic Works of the Tatar State University.* Tartu, 1977. Edition 421. pp. 9-20.

Amiragova, M.G., Kovalev, S.V., Svirskaya, R.I., The Dynamics of the Secretion of Certain Hormones during Chronic Emotional Stress. *The I.M. Sechenov Physiology Journal of the USSR.* 1979. v. 65, no. 5. pp. 694-701.

Anisimov, V.N., The Change in the Level of Biogenic Amines in the Brain of Laboratory Animals and Man during Development and Aging. *Achievements of Physiological Sciences* 1979. v. 10, no. 1. pp. 54-75.

Anokhin, P.K., *Works on the Physiology of the Functional Systems.* Moscow, 1975. 446 pp.

Anokhina, I.P., *Neurochemical Mechanisms of Psychic Illnesses.* Moscow, 1975. 320 pp.

Anokhina, I.P., *Neurohumoral Factors of Individual Resistance to Emotional Stress. Experimental and Clinical Neuroses*: (Thesis Report of the 9th Symposium of the Interbrain). Berlin, 1984. pp. 98-99.

Arefolov, V.A., Possibilities of the Complex Study of the Role of Adrenergic Mediator in the Mechanism of the Action of Psychotropic Substances. *The Neurochemical Bases of the Psychotropic Effect*. Edited by Academician of the USSR Academy of Medical Sciences A.V. Valdman. Moscow, 1982. pp. 81-94.

Azhipa, Y.I., *Nerves of the Internal Secretion Glands and Mediators in the Regulation of Endocrine Functions*. Moscow, 1981. 503 pp.

Babichev, V.N., *The Neuroendocrinology of Sex*. Moscow, 1981. 223 pp.

Bakhova, L.K., *The Influence of Dihydrotestosterone and Androstanasole on Certain Aspects of Catecholamine Exchange*: Candidate of Biological Sciences. Kharkov, 1975. 17 pp.

Balkania, G.S., Dartsmelia, V.A., The Orthograde Factor of the Development of Arterial Hypertension in Primates. *Cosmic Biology and Aviation-Cosmis Medicine*, 1984. v. 18, no. 3. pp. 14-19.

Bankova, V.V., Kucherenko, A.G., Markov, K.M., Changes in Arterial Pressure and Biosynthesis of Cortico-Steroids in Rats, in a State of Neurogenic Stress. Cardiology. 1976. v. 16, no. 9. pp. 127-128.

Bekhtereva, N.P., Kambarova, D.K., Pozdneev, V.K., *Resistant Pathological State during Illnesses of the Brain*. Leningrad, 1978. 240 pp.

Bekhtereva, N.P., Smirnov, V.M., The Cerebral Organization of Emotions. *News of the USSR Academy of Medical Sciences* 1975, no. 8. pp. 8-19.

Bekhtereva, N.P., The Health and Diseased Brain of Man. Leningrad, 1980. pp. 69-101.

Belkania, G.S., *The Functional System of Anti-Gravity*. Moscow, 1982. 288 pp.

Belova, T.I., Kvetnansky, R., Catecholamines in the Brain under Conditions of Experimental Emotional Strain. *Achievements of Physiological Sciences* 1981. v. 12, no. 2. pp. 67-90.

Belova, T.I., Vasiliev, V.N., Functional Activity of the Adrenal Cortex in Humans during Shift Work with Emotional Strain. *The I.M. Sechenov Journal of Physiology of the USSR*. 1974. v. 60, no. 3, pp. 329-333.

Berezkin, S.E., Tarasov, A.N., Excretion of Catecholamines in Urine during Hypertensive Illness and Certain of its Complications. *Clinical Medicine* 1978. v. 56, no. 10. pp. 48-53.

Blinova, N.G., Krasimova, E.I., Ksents, S.M., A Change in the Cholinergic and Adrenergic Activity of the Arterial Blood of Dogs under a Physical Load at Various Times of the Day. *Mechanisms of* *the Regulation of the Physiological Functions under Experimental Conditions.* Tomsk, 1978. pp. 26-31.

Blinova, T.S., *The Morphofunctional Nature of the Testes under Intermittent Hyperthermic Stress* (Swimming in Cold Water): Candidate of Medical Sciences Leningrad 1977. 24 pp.

Bogdanov, A.I., Filaretova, L.P., Filaretov, A.A., The Graduation of the Reaction of the Pituitary-Adreno Cortical System to the Activating and Braking Signals. *The I.M. Sechenov Physiological Journal of the USSR.* 1982. v. 86, no. 6. pp. 804-808.

Bolotova, I.A., Klingman, L.E., Ivkina, T.M., Budovsky, M.P., A Study of the Biochemical Impulses in the Biological Fluids and Erythrocites of Healthy Persons Under Experimental Intellectual-Emotional Strain. *The Biochemical Nature of Pathological Processes.* Riga, 1980. pp. 54-61.

Bolshakova, T.D., Chirkov, A.M., Startsev, V.V., The Activity of the Sympatho-Adrenal System in Monkeys during Severe and Chronic Emotional Stress. *News of the USSR Academy of Sciences. Biology Series.* 1985. no. 6. pp. 822-831.

Bolshakova, T.D., Common Principles and Methods of Hormone Research. *Cardiology Administration. v. 2: Cardiovascular System Research Methods.* Editor Academician E.I. Chazov. Moscow, 1982. pp. 473-406.

Bolshakova, T.D., Menshikov, V.V., Lukicheva, T.I., Dibobes, G.K., The Sympatho-Adrenal System in Virtually Healthy People under Emotional and Physical Loads. *Actual Problems of Biochemistry and Pathology of the Endocrine System*: Thesis Report of the National Congress of Endocrinology, Moscow, 1972. 184 pp.

Bolshakova, T.D., The Clinical Application of the Study of the Metabolism of Catecholamines. *D.I. Mendeleev Journal of National Chemistry.* 1976. v. 21, no. 2. pp. 196-203.

Budantsev, A.Y., *Monoaminergic Systems of the Brain.* Moscow, 1976. 192 pp.

Burov, Y.V., Vedernikova, N.N., Dissociation of the Endocrine and Neurotropic Properties of Peptide and Steroid Hormones. *Pharmacology and Toxicology.* 1984. v. 47, no. 2. pp. 100-105.

Butnev, V.U., Lomaya, L.P., Taranov, A.G., Goncharov, N.P., Age Differences in the Hormonal Reaction of the Adrenals and Testes to Stress Application in Monkeys. *Bulletin of Experimental Biology and Medicine.* 1980. v. 89, no. 2. pp. 157-160.

Butnev, V.U., *The Establishment of the Endocrine Function of the Adrenals and the Gonads in the Ontogenesis of Papio Hamadryus and Growth Characteristics of the Hormonal Reaction to Stress*: Candidate of Medical Sciences. Leningrad, 1980. 16 pp.

Butovskaya, M. L. , Deryagina M.A., Chirkov, A.M. Stratsev V. G. The effect of Stress Factors on the Behavior of Monkeys. II.

Hormonal Indicators and their Connection with Behavior in the Presence of Models of Emotional Stress on Hamadryad Baboons. *Biological Sciences*. 1986. no. 2. pp. 59-64.

Butovskaya, M.L., Deryagina, M.A., Chalyan, V.G., Chirkov, A.M., Startsev, V.G., The Ethological in Psychopharmacological Research. *Problems of Providing Monkeys for Medical-Biological Research and the Principles of the Use of Monkeys in Experiments*: (Materials of the National Conference). Sukhumi, 1983. pp. 39-40.

Butovskaya, M.L., Deryagina, M.A., Chirkov, A.M., Startsev, V.G., Chalyan, V.G., The Influence of Stressogenic Factors on the Behavior of Monkeys. I. The Behavior of Monkeys under Conditions of Severe Emotional Stress. *Biological Sciences*. 1985. no. 6. pp. 67-74.

Cherkovich, E.M., Lapin, B.A., Cardiovascular Pathology under Experimental Neurosis on Monkeys. *Physiology and Pathology of the Cortico-Viscera Relationship*.Leningrad, 1978. pp. 174-182.

Cherkovich, G.M. Fufacheva, A.A., *Heart Components of Emotional Reactions of Monkeys. Biology and Acclimatization of Monkeys.* Moscow, 1973. pp. 114-117.

Cherkovich, G.M., Rutkay-Nedetskii, I., Fufacheva, A.A., *Monkeys-- Object of Studies on Physiological and Pathological Emotions. Problems of Providing Monkeys for Medical-Biological Research and Principles of Utilizing Monkeys in Experiments.* Sukhumi, 1983. pp.86-87.

Chirikov, A.M., Chirkova, S.K., Startsev, V.G., Tsulaya, M.G., Role of Monoamino Systems and Steriod Producing Glands in the Development of Emotional Stress and Pathological Arterial hypertonia in Monkeys. *Cardiology*. 1986z. Vol. 26, no., 1 pp. 49-54.

Chirikova, S.K., Chirikov, A.M., Voyt, I.S., Neurohormonal Mechanisms of Resistance and Behavior of Monkeys under Active Psychogenic and Alcohol Stress. *Neurohormonal Regulations of immune Homeostasis*. Leningrad, 1986. pp. 256-257.

Chirikova, S.K., Startev, V.G., Katsiya, G.V., *Study of Antistress Mechanisms; Effect of New Psychotropic Preparations on Models of Pathological Condition when Searching for Biological Active Preparations.* Moscow, 1983. Chapter 2, pp. 6-7.

Chirkov, A. M., Chirkova, S.K., Tsuliya, M.G., Boht, I.S., *Neuroendocrinal Mechanisms of Neurogenetic Arterial Hypertonia in Monkeys.* Excerpt of the Lecture of the Twelfth Conference of the Union of Physiology and Pathology of the Cortical-Visceral Correlation, Dedicated to the 100th year Anniversary of the Opening of the Academy of K.M. Bikov. Leningrad, 1986b. pp.286.

Chirkov, A.M. Katsiya, G.V., Methodological Approach to the Study of the Effects of Biological Activities of the Enviroment on Endocrinal Function of the Testicles. *Utilization of Models of Pathological Conditions for the Search for Biologically Activated Preparations.*Moscow, 1983. pp. 96-97.

Chirkov, A.M. Startsev, V.G., Krilov, S.S., *The Effect of the Pituitary on the Function of the Neuroendocrinal and Cardio-Vascular System of Hamadryas Baboons in Situations of Stress.* Leningrad, 1980. pp. 185-186.

Chirkov, A.M., *Functional Activity of the Sympathetic Adrenal System and Steriod Producing Glands under Stressful Activity of Monkeys*:Abstract of Dissertation for Candidacy of Medical Sciences. Leningrad, 1984, pp. 26.

Chirkov, A.M., Goncharov, N.P., Effect of Many Short Neuroemotional Stresses on the Cardiovasculal and Neuroendocrinal Systems of Monkeys. *Biological Characteristics of Laboratory Animals and Extrapolations to Human Experimental Data.* Moscow, 1980b. pp. 187-189.

Chirkov, A.M., Goncharov, N.P., *Role of Neuroendocrinal Changes in the Development of Arterial Hypertonia of Monkeys.* Material for the Second Congress of the Union of Endocrinologists, Leningrad, 1980a. pp. 185.

Chirkov, A.M., Krilov,S.S., Katsiya, G.V., Effects of the Leuliberinia on Exertion of Cathecholemines of Hamadryas Baboons in Situations of Stress. *Perspectives of Bioorganic Chemistry in the Creation of New Medical Preparations.* Riga, 1982. pp. 71.

Chirkova, S.K., *Effects of Extreme and Frequent Immobilizational Stress on the Exertion of Catecholemines and their Precusors in Immature Monkeys. Adaptation to Experimental Conditions.* Excerpt of the Lecture of the Sixth Conference of the Union of Ecological Physiology. Siktivkar, 1982. Vol. 4. pp. 99-100.

Chirunov, V.S., Vasilev, V.N., *Neurosis, Neurosis-like Situations and Sympathetic Adrenal System: Objective Diagnosis, Therapy and Prognosis.* Moscow, 1984. pp. 192.

Davidova, N.A., *The Study of the Activity of the Sympathetic-Adrenal System of Mice under Immobilizational Stress. Stress and Adaptation*: Text of the All Union symposium. Kishenev, 1978. pp. 95.

Deryagina, M.A., Butovskaya, M.L., Stratsev, V. G., Chirkov A.M., Chalyan V.G., *Ethological Analysis of the Behavior of Monkeys in Medical-Biological Research.* Biol.Sciences. 1984. no. 7 pp. 57-61.

Dimitrov, D. J., *Chronic Gonadotropism of Man.* Moscow. 1979. pp. 143.

Djalagonia Sh.L., *Experimental Neuroses of Monkeys*: Abstract of the Dissertaion of Doctorate of Medical Sciences. Leningrad. 1979. pp. 33.

Efremova, G.V., Seasonally Changes Circadian Rythyms of Activity of the Cortical Adrenal Glands and the Particulars Morphofunctionality of Adaptation of Mice Activities. *Physiological Aspects of Adaptaion of Man and Animals.* Kererovo, 1978. pp. 35-46.

Fedorov, B.M., *Emotions and Heart Activity.* Moscow, 1977. pp. 216.

Filaretov, A.A., *Nervous Regulation of the Hypophysio-Adrenalo-Cortical System*. Leningrad, 1979. pp. 144.
Firsov, L.A., *I. P. Pavlov and Experimental Primatology*. Leningrad, 1982. pp. 156.
Fokin, A.S., *Changes in the Lipids Exchange under Longterm Functional Stress of the Nervous System*: Abstract of the Dissertation for Doctorate of Medical Sciences. Moscow, 1981. pp. 32.
Fridman, E.P., Biological Prerequisites and Collective Characteristics of Medical Research on Monkeys. *Bulletin of the Academy of Medical Sciences of the USSR* 1977. no. 8, pp. 72-80.
Furdiy, F.I., Babare, G.M. Guragata, E.N., Krivosheev, O.I., Marin, L.P.,Nikitovich, S.N., Pryanishnikova, O.V., Tofan, S.S., Fuduy, M.F., Khaydarliu, S.Kh., Shvareva, N.V., Functions of Certain Internally Secreting Glands under the Influence of Extreme Stress on the Organism. *Stress and its Pathological Mechanisms*. Kishenev, 1973. pp. 43-45.
Ganelina, I. E., Borisova, I. U., Twenty-four Hour Rhythms, Capabilities, Activity (level) of the Sympathetic-Adrenal Systems and Coronary Myocardia. *Physiology of Man*. 1983. Vol. 9. no.2 pp. 249-256.
Garkovi, L. X., Kvakina, E. B., Ukolova, M. A., *Adaptive Reactions and Resistance of an Organism*. Rostov on the Don, 1977. pp. 120.
Gerasimova, E. N., Glazunov, I.S. , Bodrova, E.A., Suchkova, S. H., Zikova, V.P., Vurlutsii, G. I., Perova, N.V., Chernishova, N.P., Zadoya, A.A., Changes in the Number of Hormones in Blood Plasma and Risk Factors for Isometric Diseases of the Heart for Men between the Ages of 40-59. *Ques. of Med. Chemistry*. 1978. Vol. 24. Abstract 5. pp. 657-666.
Gladkova, A. I., Ozerova, M. R., Bondarenko, L.A., Adrenergic Regulation of Male Sex Glands with the Application of Low Temperature Influence on Prostrate Gland. *Endocrinology*. Kiev, 1982, Abstract 1. pp. 35-39.
Goncharov, N.P., Antonichev, A. V., Katsiya, G.V., Butnev, V. Y. Taranov, A. G., Lanin, B. A., Character of Change in leveling the Steroidal Hormones and their Preceding in Peripheral Blood of Baboons under Conditions of Stress. *Biol. Exp. of Biology and Medicine*. 1978z. Vol. 85, no. 3, pp. 270-274.
Goncharov, N.P., Butnev, V. Y., Characteristic Endrinocrinal Functions of Adrenal Glands of Hamadryad Baboons under Conditions of Metopionic Testing. *Problems of Endocrinology* 1984. Vol. 30, no., 2. pp. 81-84.
Goncharov, N.P., Chekan, S., Antonichev, A. V., Katiya, G. V., Butev, V.Y., Disfaluzi, E., Radioimmuizational method to Determine II Steroids in Small Quantities of Monkey Plasma. *Quest. of Medical Chemistry*. 1979. Vol.25. no.1. pp. 92-97.

Goncharov, N.P., *Functions of the Cortex of the Suprarenal Glands of Lower Monkeys in Normal and Certain Pathological Situations*: Abstract from Dissertaton for the Doctorate of Medical Sciences. Leningrad, 1971, pp. 43.

Goncharov, N.P., Katsiya, G. V., Antoninchev, A. V., Butnev, V. Y., Utilization of Radioimmunizational Methods for the Determination of Steroids with the Application of EVM for the Study of Hormonal Function of the Adrenals and Gonads of Primates. *Models of Pathological Condition of Man*. Moscow, 1977b. Vol. 2. pp. 58-68.

Goncharov, N.P., Katsiya, G.V., Aso T., Chekan, S., Disfauzi, E., The Role of Adrenal Glands and Testicles in the Formation of Pool of Sexual Steriods in Peripheral Blood of Males Monkeys. *Prob. Endocrinology*. 1976b. Vol. 24. no.1. pp. 98-102.

Grigoreva, Z.E., Sympathetic-Adrenal System under Chronic Cardiovascular Deficencies of Patients with Hypertonic Disease. *Physicians Profession*. 1978 no. 9. pp. 29-32.

Gubachov, Y. M., Stabrovski, E. M., *Clinical-Physiological Basis for Pychosomatic Correlations*. Leningrad. 1981. pp. 216.

Ibatov, A. N., Clinical Utilization of Pituitary Serum in the Practice of Psychology. *Preservation of Health of Kazakhstan*. 1979. no. 2 pp. 46-47.

Ignatov, Y. D., Galustyan, G.Z. Andreev, B.V., The Role of GAMK-ergical Mechanisms in Stress-Protective Effects of Benzodiazepine Group of Tranquilizers. *Neurological Basis of Psychotropic Effects*. Under the Editing of Academician of the Academy of Medical Sciences of the USSR by A.V. Valdman. Moscow, 1982, pp. 118-126.

Kassil, G. N., Matlina, E. Sh., Sympathetic-Adrenal System under Stress. Material from the Symposium of All States, *"Stress and its Pathological Mechanisms"*. Kishinev, 1973. pp. 24-26.

Kassil, G. N., Matlina, E. Sh., Vasilev, V. N., Kikolov, A. I., The Effect of Stressful Mental Work during Day and Nocturnal Hours on the Excretion of Cathecholaminics in Urine.. *Physiological Journal of the USSR*. 1973. Vol. 59, no. 8. pp.1151-1157.

Kassil, G.H., Humoral-Hormonal Mechanisms for the Adaptation of the Organism to Athletic Activities. *Physiology of Man* 1975. vol. 1. no. 6. pp. 1032-1047.

Kassil, G.N., Some General Conformities for the Regulation of the Sympathetic-Adrenal System during The Adaptation of the Organism to Physical Exertion. *Nervous and Endocrinal Mechanisms of Stress*. Kishinev, 1980. pp. 122-135.

Katsiya, G.V., Chirkov, A. M., Goncharov, N. P., Endocrine Function of Seminal Fluid of Hamadryan Baboons (Papio Hamadryas) in Situations of Chronic Emotional Stress. *Bul. Experiments of Biology and Medicine*. 1984b. Vol. 97. no. 3. pp. 285-287.

Katsiya, G.V., Chirkov, A. M., Goncharov, N. P., The Effect of Luliberin in Chronic Gonadaltropism on the Level of Luteinic Hormone and Testoteronein the Blood of Monkeys in Situations of Extreme Stress. *Prob. Endocrinology.* 1984a. Vol. 30. no. 1. pp. 73-76.
Katsnelson, Z.S., Stavrovskii, E.M., *Histolic and Biochemical Chromaffinal Fabric of the Pituitary.* Leningrad. 1975. pp. 224.
Kazin, E. M., *Circadian Rhythms of Adrenal-Cortal Activity During the Adaptation of the Organism to Factors of the Surrounding Environment*: Abstract of the Dissertation for Doctorate of Medical Sciences. Leningrad. 1982. pp. 37.
Khasabov, G.A., *Neurophysiological Correlation of the Cortex of the Large Hemisphere of Primates.* Moscow, 1978. pp. 182.
Khaydarlinu,S.Kh, Godukhin, O.V., Budatsev, A.U., Effect of Glucocortiod Dissemination on The Liberation and Catabilism of H3 Dopamine in Tail Nucleus of Rats. *Neurochemistry,.* 1983. Vol. 2 No. 2 pp. 130-137.
Khomulo, P.S., *Emotional Stress and Arteriolosclerosis.* Leningrad, 1982. 152. pp.
Kitaev-Smik, L. A., *Pychological Stress.* Moscow. 1983. pp. 368.
Komissarenko, V.P., Rezikov, A.G., *Inhibitors of the Function of the Cortalpituitary Gland.* Kiev, 1972. pp. 374.
Korkach, V.I., *Role of AKTG on Glucocorticals in the Regulation of Heart Activities.* Leningrad 1979. pp. 152.
Korneva, E.A., *Evolution of Reflexive Regulation of Heart Activity.* Leningrad. 1965. pp. 250.
Korneva, E.A., Klimenko, V.M. Shkhinek, E. K., *Neuro-hormonal Provisions of the Immuno-Homeostase.* Leningrad. 1978. pp. 176.
Korobova, A. A., Matlina, E. Sh., Vasilev, V.N., Balashov, N.N., Galimov, S.D., Ogoltsov, I.G., Smirnova, T.A., Condition of the Sympathetic-Adrenal Systems and Some Indications of Change in Dynamic Processes of Training Cycles of Athletes. *Lab. Work.* 1977. no. 8. pp. 469-478.
Kozlovksaya, M.M., Klusha, V. E., Bondarenko, N.A., Comparison of Psychotropic and neurochemical Activity of Short Peptides. *Neurochemical Basis of Psychotropic Effects.* Under the Editing of Academician of the Academy of Medical Sciences of A. V. Valdman. Moscow. 1982. pp. 95-105.
Krachun, G.P., Role of Mediators of the Sympathetic Conductors in the Central Neuro-Hormonal Mechanisms for Regulation of the AKTG-Corticalsteriodal System Functions. *Vest. Academy of Medical Sciences.* 1977. no. 9. pp. 85-91.
Kruglikov, R. I., *Neurochemical Mechanisms for Learning and Memory.* Moscow, 1981. pp. 211.
Ksents, S. M., Blinova, N. G., Dynamics of Some Indicators of the Sympathetic Adreanal System Under Physical Exertion During

Various Times of the Day. *Problems of Experimental and Morphology and Genetics.* Kemerovo, 1976. pp. 44-47.

Kuksova, M. I., Functional Characteristics of Hemotogenetic Monkeys. *Models of Pathological Conditions of Man.* Moscow. 1977. Chap. 2 pp. 40-46.

Kulagin, V.K.,, Davidova, V. V., Clinical-Experimental Data on the Reaction of the hypothalmic-hypophysical-Adrenal Systems During Extreme Mechanical (Physical) Trauma. Tez. Symposium of the whole union *"Stress and Adaptation".* Kishinev, 1978. pp. 116-117.

Kvetnanskii, R., Belova, T.I. Oprshalova, Z., Ponets, I., Iindra, A., Dushkin, V.A., Cathecholistic Blood Plasma of Rats From the Line of August and Vistar under Emotional Stress. *Physio. Journal of the USSR in the name of I. M. Sechenov.* 1981. Vol. 67, no. 4. pp. 516-523.

Lapin, B. A. Chirkov, A.M., Chirikova, S.K., Tsulaya, M.G. , Chalyan, V.G., Deryagina, M.A., Changes in the Behavior and Function of the Neuroendocrinal System in Situations of Emotional Stress. *Emotions and Stress: A Systematic Approach.* Moscow, 1984. pp. 178-179.

Lapin, B.A. Chirkov, A.M., Chirikova, Stress Factors of the Surrounding Enviroment as a Condition for the Development of Neurosis and Somatic Disorders of Monkeys. *Actual Problems of Stress.* Kishenev, 1976, pp. 151-153.

Lapin, B.A., A Few Summations and Perspectives on the Utilization of Monkeys in Medical-Biological Experiments. *Vest. Academy of Medical Sciences of the USSR* 1977. no. 8. pp. 3-12.

Le Thu Lien, Chernositov, A.V., Electrophysiological Analysis Androgenetical Mechanisms of the Hypothalmic Reglutation of Male Sex Glands. *Bulletin of Experimental Biology and Medicine.* 1983. Vol. 95, no. 3. pp. 18-19.

Lemondjava, N.I., Djalagoniya, Sh.L., The Effect of Daily Blood Draws on the Activity of Creatinephosphous, AST and ALT in Blood Serum of Monkeys. Questions of Medical Chemistry, 1981. Vol. 27., no. 6. pp. 64-67.

Magakyan, G.O., Experimental Changes in the Pathology of Arterial Hypertonia and Isometric Disease of the Heart on Monkeys. *Vest. Academy of Medical Sciences of the USSR, 1977.* no. 8. pp. 20-24.

Malishenko, N.M., Activity of the Corticosteriods on the Hypothalmic-Reticuloendothelial Education of the Advanced Brain. *Successes of Physiological Sciences,* 1976. Vol. 7. no. 2. pp.88-115.

Manukhin, B.N., Pavliva, V.I., Putinteva, T.G., Volina, E.V., Berdisheva, L.V., Kurbanova, G.D., Selivanova, G.P., Meerson, F.Z., Functional Condition of the Sympathetic-Adrenal System of

Rats under Emotional-Traumatic Stress. *Physiological Journal of the USSR named for I.M. Sechenov,* 1981. Vol. 67, no. 8. pp. 1182-1188.

Marin, L.P., Guragata, E.H., Role of the Hypophysis, Pituitary and Sex Organs in the Development of Endurance of an Organism to Extreme Conditions. Tez. Symposium of the whole Union. *"Stress and Adaptation",* Kishenev, 1978. pp. 123-124.

Markaryan, R.L., Samsons, V.M., Babicheva, V.N., The Influence Mono-amines on the Content of Leuliberia in Different Sections of the Hypothalamus of Male Rats. *Problems of Endocrinology,* 1983. Vol. 29, no. 2 pp. 60-64.

Markel, A.L., The Role of Catecholamines in the Development of Spontaneous Arterial Hypertension in Rats with a History of Hypertension. (SHR-Spontaneously Hypertensive Rats) *Successes of Physiological Sciences.,* 1983, Vol. 14, no. 1 pp. 67-84.

Markel,A.L., Kazin, E.M., Lurye, S.B., Naumenko, E.V., Influence of Stress in Early Ontogenesis on the Circadian Rhythm of the Corticosteriodal Functions of Rats. Ontogenesis. 1981. Vol. 12, no. 3. pp. 257-265.

Martirosyan, V.V., Pokrovskaya, E.A., Eremina, S.A., Polyak, A.L., *Curative Effects of the Pituitary in the Presence of Hypothalmic Vegetative Vascular Paroksism.* Material from the 4th All Union Conference on Physiological Vegetative Nervous System. Yerevan, 1976. pp. 205

Matushov, M.I., Hyperphysio-Adrenal System and Stress. *Hyperphysio-Adrenal System and the Brain.* Leningrad, 1976. pp. 192-204.

Meerson, F.Z., *Adaptation, Stress and Prevention.* Moscow, 1981. pp. 277

Meerson, F.Z., *Pathogensis and Indicators of Stress and Isometric Injury of the Heart.* Moscow, 1984. pp. 272.

Menshikov, B.B., Bolshakova, T.D., Principles of Research and Nature of Reactionary Changes of Catecholemines of People in Situations of Stress and Adaptations. Tez. symposium of the All-Union *"Stress and Adaptation".* Kishinev, 1978. pp. 35-36.

Menshikov, V.V., Berezin, F.B., Bolshakova, T.D. Tyabenkova, V.F., Solovenchuk, L.L., Dibobes, G.K., Laneeva, A.I., Experiments with Catecholamines in Urine from Heathy People Situated on the Northwest Edge of the USSR. *Lab. work,* 1977. no. 9, pp.539-544.

Mezentseva, L. N., *Role of Individual Particulars of Change of Biogenetic Amines in the Stability to the Development of Pathological Reactions of Stress.* Abstract of the Dissertation for Doctorate of Biological Sciences. Moscow, 1982. pp.17.

Mikhin, V. Kh., Agzamkhadsaev, T.S., Sharipov, T.T., et al., *Sympathetic Adrenal Reactions of Children as a Reaction to Stress Indicative of Surgical Pathology.* tp. Second Moscow Medical Institute. 1978. Vol. 114, Excerpt 2. pp. 91-98.

Miloslavskii,J. M., Menshikov, V.V., Bolshakova, T.D., *Pituitary Arterial Hypertonia.* Moscow. 1971. pp. 260

Mustafin, M. G., Sitdikov, F.G., Ages where the Sympathetic Adrenal System is Particularly Reactive under the limits for Physical Exertion. *Physiology and Biochemistry Mediating Processes*: (Tez. of the Lecture of the Third All Union Conference.) Moscow, 1980. pp. 147.

Naumenko E.V., Digalo, N.N., Brain Noradrenaline Mechanisms and Emotional Stress in Adults Rats after Prenatal Hydrocortisone Treatment. *Catecholamines and Stress.* 2nd Intern. Symp., Sept. 12-16, 1979, Smolenice Castle, USSR, Abstracts. P. 54.

Naumenko, E.V., Biogenetic Amines in Regulation of the Mechanism of Reciprical Negative Connection of the Hyperthalmic-Hypophysical-Sex Complex. *Mechanism of Hormonal Regulation and their Role in the Reciprocal Connection in the Development of Homeostasis.* Moscow, 1981. pp. 140-151.

Naumenko, E.V., Popova, N.K., *Serotonin and Melatonin in Regulation of the Endocrinal System.* Novosibirsk, 1975. pp. 218.

Nazdrachev, A.D., *Corticosteriods and the Sympathetic Nervous System: (Electrophysiological Study of the Functions of the Peripheral Area)* Leningrad, 1969. pp. 117

Nazdrachev, A.D., Pushkarov, U.P., *Characteristics of Medial Transformation.* Leningrad, 1980. pp. 230.

Obibok, V.N., *Amount of Catecholamines in Various Sections of the Myocardia of a Man in Ontogenesis Suffering from Certain Heart-Vascular Diseases.* Abstract for the Dissertation for Doctorate of Medical Sciences. Leningrad, 1981. pp. 22.

Obut, G.A. Gizatulin, Z. J., Jacobson, G.S., *Two Forms of Neuroendocrinal Reactions of the Organism to Often Repeated Extreme Pressures.* Material from the Second All Union Congress of Endocrinologists. Leningrad., 1980. pp.334-335.

Orbeli, L.A., Adaptive-trophic Role of the Sympathetic Nervous Systems, Cerebellum and Higher Nervous Activity. *Physiology Journal of the USSR in the name of I.M. Sechenov.* 1949. Vol. 35. pp. 594-595.

Osinskaya, V.O., Rasin, M.S., Utevskii, A. M., Certain Biochemical Mechanisms of Interaction and Function of the Catecholamines and corticosteriods. *Physiological, Biochemical and Pathological Endocrine Systems.* Kiev, 1975. Excerpt 5. pp. 3-11.

Panafilova, O. V., Palchikova, N.A., *Of Mechanisms Interacting with the Sympathetic-Adrenal and Hypothalmic-Hypophysin-Adrenocortical System in Regulation of Metobolic Processes.* Material of the 2nd meeting of the Union of Endocrinologists. Leningrad, 1980, pp.341-342.

Panin, L.E., *Biochemical Mechanisms of Stress.* Novosibirsk, 1983. pp.233.
Paskaleno, A.N., Zozulya ,R. N., Davidov, V.V., The Effect of Pituitary and Leftmepromazine on Certain Evidence of Neuroendocrinal Regulations. *Neuropharmocology: (New Preparations in Neurology)* Leningrad, 1980. pp. 140-141.
Pau, A.U., *The Study of Roles of the Sympathetic Adrenal Systems in Adaptation of Trained and Untrained Organisms under Physical Exertion*: Dissertation for Doctorate of Medical Sciences. Tarta, 1975. pp.342.
Popova, N.K., Koryakina, L. A., *Of the Role of Catecholemines in Various Forms of Stressful Reactions.* Materials form the Second Meeting of the Union of Endocrinologists. Leningrad, 19809. pp. 350-351.
Poppai, M., Vakhtel, Gekht, K., Shlegel, T., Zass, R, Lipka, K., *The Significance of the Phase of Sensitivity of the Circadian Rhythm for the Development of Experimental Neurosis.* Thesis of a Lecture of the Ninth Symposium of the Interbrain). Berlin, 1984. pp. 30-31.
Poshivalov, V.P., Neuropharmocological Aspects of Anti-Aggressive and Rheo-Socialistic Effects. *Neurochemical Basis for Psychotropic Effects.* Under the Editing of Academician of the Academy of Medical Sciences of the USSR. A. V. Valdman. Moscow., 1982. pp.106-117.
Repin, U.M., Startsev, V. G., Mechanisms of Selected for the Defeat of the Cardiovascular Systems under Psychoemotional Stress. *Bulletin of the Academy of Medical Sciences of the USSR.* 1975. no. 8, pp. 71-78.
Robu, A.I., *Correlation of the Endocrinal Complexes under Stress.* Kishenev, 1982. pp. 207.
Rotenberg, V. S., Arshavskii, V.V., *Exploratory Activity and Adaptation.* Moscow, 1984. pp. 193.
Rubanova, N.A., Riman, R.S., The Role of Biological Amines of the Brain in Sudden Adaptation to Hypocia. *Dehydrogenation in the Norms of Pathology.* Gorki, 1980. pp. 57-62.
Rugal, V.I., *Morphology of Male Sex Glands under Stress*: Abstract from the Dissertation for Candidacy of Medical Sciences. Leningrad, 1977. pp. 20.
Sapronov, N.S., *Mechanisms of the Central Regulation of the Hypophysio-Pituitary System (Physiological and Pharmocological Analysis)*: Abstract of the Dissertation Doctorate of Medical Sciences. Leningrad., 1980. pp. 42.
Savina, E.A., Panikova, A.S., Kabintskaya, O. E., Zagorskaya, E.A., Lubarskaya, I.I., Certain Reactions of the Cortex of the Pituitary on Immobilizational Stress of Intact Animals and on Background of Hypokenesis. *Archives of Anatomy, Hystology and Embyology.* 1980. vol. 42. Excerpt 9. pp. 66-72.

Shalyapina, V.G., Rakitskaya, V.V., *Effect of the Corticosteriods on the Substance and Metobolic Conversion of the Catecholemines in the Brain of Animals*. Hypophysio-Adrenal System and the Brain. Leningrad, 1976. pp. 67-86.
Shalyapina, V.G., Role of the Catecholemines of the Brain in the Regulation of the Hypophysio-Adrenal System. *Hypophysio-Adrenal System and the Brain*. Leningrad, 1976. pp. 46-66.
Shalyapina,V.G., Brain Matter of the Pituitary. *Physiology of the Endocrine System*. Leningrad, 1979. pp. 325-340.
Shorin, U.P., Certain Endocrinal Mechanisms of Regulation of the Restorative Processe. Materails of the Second Symposium "*Adaptation and Adaptogenesis*" Vladivostok, 1977. pp. 52-53.
Shorin, U.P., Obut, T.A., *Changes in the Regulation of Mechanisms of the Hypothalmic-Hypophysio-Pituitary Systems in the Process of Adaptation to Stress. Adaptation and Problems of General Pathology.* Excerpt from the Lecture from the Conference of the Union. Novosibirsk. 1973. Vol. 2 pp. 120.
Shorin, U.P., Obut, T.A., Ribalova, O.N., *Activity of the Hypothalmic-Hypophysio-Pituitary Systems of Male Rats after Extreme and Chronic Stress*. Selected from the Union of Academy of Sciences. Biology Series. 1975. Excerpt 3, no.15. pp.149-151.
Shurigin, D.Y., *Interrelationship of the Anabolic and Catebolic Hormones under Traumatic and Radiation Illnesses*. Materials from the Second Conference of the Union of Endocrinologists. Leningrad., 1980. pp.666-668.
Simonov, P.V., *The Emotional Brain*. Moscow., 1981 pp. 215.
Sokolov, E.M., Belova, E.V., *Emotions and Pathology of the Heart*. Moscow. 1983. pp. 309.
Stabrovskii, E.m., Egorkova, A.S., Shpanskaya, L.S., Yampolskaya, L.I., Influence of Chemical Breakup of Monoaminio Terminals of the Hypothalmus on the Function of the Cortex of the Pituitary, Thyroid Glands and Intrachromaffine Systems. *Problems of Endocrinology*. 1981, Vol. 27, no. 2, pp. 62-66.
Stanishevskaya, A.V., Mezentseva, L.H., Influence of Certain Psychopharmocological Preparations on Adaptation in Situations of Stress. *Pharmacology and Toxicology*. 1977. Vol. 40, no. 1 pp. 9-12.
Startsev, V.G., Chirkov, A.M., Longterm Effects of Cardiopathological Emotional Stress in Hamadryas Baboons. Models of Pathological Human Conditions. Moscow, 1977. Chapter 2 pp. 180-188.
Startsev, V.G., Mamamtavrishvili, S.K., Chirkova, S.K., Chirkov, A.M., Dzholiya, T.M., Kondakchyan, A.V., Antistressful and anti-neurological Activity of Psychotropic Preparation in Experiments on Monkeys. *Bulletin of the Academy of Medical Sciences of the USSR*, 1986. no. 3, pp. 30-33.

Startsev, V.G., *Models of Human Neurogenic Illnesses in Experiments on Monkeys.* Moscow, 1971. Vol. 40, no. 1 pp. 9-12.
Startsev, V.G., *Neurogenical Gastric Achilia in Monkeys.* Leningrad, 1972. pp. 200.
Startsev, V.G., *Primate Models of Human Neurogenic Disorder Disorders.* Hillsdale, New Jersey: Lawrence Erlbaum Associates Publishers, 1976. pp. 198.
Startsev, V.G., Problems of Discrimination of Injured Functional Systems under Emotional Stress and Neurosis. *Bulletin of the Academy of Medical Science of the USSR.* 1977. no. 8. pp. 32-40.
Sudakov, K.V. et al. Sudakov, K.V., Anokhina,I.P., Belova, T.I., Boliakin, V.I., Ivanova, T.M., Skotzelas, J.G., Jumatov, E.A., Catecholamine Content in the Brain of Rats with Different Resistance to Emotional Stress. *Cathecholamines and Stress.* 2nd International Symposium, Sept. 12-16, Smolenice Castle, USSR, 1979 Abstracts. pp. 79.
Sudakov, K.V., *Emotional Stress and Adaptive Hypertension: (Survey of Experimental Data).* Moscow, 1976. pp. 116.
Sudakov, K.V., *Systematic Mechanisms of Emotional Stress.* Moscow, 1981. pp. 229.
Sudakov, K.V., To the Pathogenetic destruction of Heart Activity in Situations on Emotional Stress. Path. *Physiology.* 1979. Excerpt 3 pp. 16-21.
Taranov, A.G., *Biological Rhythms of Hormonal Functions of Pituitaries and Sex Glands of Male Hamadryas Baboons and the Characteristic Changes under Stressful Situations.* Abstract of Dissertation for the Candidacy of Biological Sciences. Leningrad, 1981. pp. 29.
Taranov, A.G., Goncharov, N.P., Characteristic Circadian Rhythms and the Contents of the Corticosteriods and their Precursors in the Blood of Hamadryas Baboons in Correlation with the Initial Functional Condition of the Hypo-physio Cortex Pituitary System. *Biological Experiments of Biology and Medicine,* 1981. Vol. 41, no.2 pp. 219-222.
Tatojan, S.K., Cherovich, G.M., The Heart Rate in Monkeys (Baboons and Macaques) in Different Physiological States Recorded by Radiotelemetry. *Folia Primat.*, 1972. Vol. 17, n.4. pp.255-266.
Tavadyan, D.S., Goncharov, N.P., Functional Systems of the Hypothalamus Hypophysio-Gonads of Macacus in Situation of Longterm Hypoknesis. *Journal of Evolutionary Biochemistry and Physiology.* 1981. Vol., no.2. pp. 216-217.
Tavadyan, D.S., *Hormonal Functions of Pituitary and Sex Glands of Male Macacus Rhesus under Longterm Hypokenesis*: Abstract from Dissertation for Candidacy of Medical Sciences. Leningrad, 1981. pp.20

Tigranyan, R.A., *Metabolic Aspects of Problems of Stress in Space Flights*. Moscow, 1985. pp. 222.
Tikh, N.A., *Prehistoric Society: A Comparative-Psychological Research*. Leningrad, 1970. pp.311
Tokar, V.I., The Effect of Chrionic Gonadaltropism on Exertions of Estrogen in Healthy Men and Those Ill with Fluorosis. *Problems of Endocrinology* 1980. Vol. 26, no. 1 pp. 34-38.
Tsulaya, M.G. , Chirkov, A.M. Chirikova, S.K., Stratsev, V.G., *Activity of Sympathetic Adrenal Systems of Monkeys in Situations of Extreme Emotional Stress*. 1984. Series of Biology Vol.10, No.5 pp.293-299.
Tsulaya, M.G., Chirikova, S.K., Startsev, V.G., Butnev, V.U., *Neuroendocrinal Mechanisms of Emotional Stress of Immature Monkeys*. 1984b. Vol. 116, no. 3 pp. 653-656.
Tsulaya, M.G., *Daily and Reoccurring Details of Hormonal Developments of Emotional Stress of Male Hamadryas Baboons*: Abstract of the Dissertation for Candidacy of Medical Sciences. Leningrad, 1985. pp. 28.
Uchaina, R.V., *Contents of Corticosteriods, Catecholemines and Serotonin of Healthy People in Connection with the Effect of Certain Factors of the External Enviroment*: Abstract of the Dissertation for Candidacy of Biological Sciences. Moscow, 1983, pp. 20.
Udaev, N.A., Afinogenova, S.A., Goncharov, N.P., Katsiya, G.V., Ways of Biosynthesis of Corticosteriods in the Pituitaries of Macaques Rhesus Monkeys. *Problems of Endocrinology.* 1977. Vol. 23, no. 5. pp. 92-96.
Udaev, N.A., Afinogenova, S.A., Krekhova, M.A., Corticosteriods. *Biochemical Hormones and Hormonal Regulation.* Edited by N.A. Udaev. Moscow. 1976. pp. 171-227.
Urmancheeva, T.G., Bauman, Kh., Khasabova, V.A., Martin, G., Bolter, F., Emotional Stress of Lower Monkeys. *Journal of Higher Nervous Activity named after I.P. Pavlov,* 1977a. Vol. 27. Excerpt 1 pp. 335-337.
Urmancheeva, T.G., Bauman, Kh., Khasabova, V.A., Martin. G., and et al., Models of Nervous Disorders in Monkeys in Situation of Extreme Exertion on Higher Nervous Activity. *Bulletin of the Academy of Medical Sciences of the USSR.* 1977v. no. 8. pp. 65-72
Urmancheeva, T.G., Bauman, Kh., Khasabova, V.A., Martin. G., Bolter, F., Neurological Disorders of Monkeys under Extreme Exertion for Analytical Processes. *Journal of Higher Nervous Activity in the name of I. P. Pavlov,* 1977b, Vol.27, Excerpt 1 pp. 64-72.
Urmancheeva, T.G., Functional Characteristics of Hypocamp in Lower Monkeys. *Journal of Higher Nervous Activity named after I. P. Pavlov.* 1972, Vol. 22, no.6. pp.1234-1241.

Urmancheeva, T.G., Khasabova, V.A., Bauman, Kh, Bolter, F.. Mechanical Development of Pituitary Disorders of Monkeys. *Journal of Higher Nervous Activity. in the name of I.P. Pavlov.* 1977g. Vol. 27. Excerpt 2. pp. 338-341.

Urmatov, E.A., Peptide-Neuromediating Mechanism that Endures Emotional Stress. *Stress and Psychic Pathology*: (Collection of Scientific Works) Moscow., 1983. pp. 7-12.

Utevskii, A.M. Osinskaya, V.O., Changes of Cathecholemines and Certain Mechanisms of Adaptation. *News about Hormones and Mechanisms of their Action.* Kiev, 1977. pp. 123-133.

Utevskii, A.M., Gaysinskaya, M.U., Effect of ACTG on the Restitution of Catecholemines in Different Organs of Animal. *Physiology, Biochemistry and Pathology of Endocrine Systems.* Kiev, 1971. Excerpt 1, pp. 27-30.

Utevskii, A.M., Rasin, M. S., Catecholemines and Corticosteriods: (Molecular Aspects of the Interrelationship of the Two Basis Adaptive Systems). *Successes of Modern Biology.* 1972. Vol. 73, Excerpt 3 pp. 323-341.

Utkin, I.A., Dynamic Processes in the Organism in Connection with the Reaction on to the *Novelty of the External Environment. Physiology and Pathology of the Higher Nervous Activity.* Sukhumi, 1960, pp. 7-16.

Valdman, A. V., Kozlovskaya, M. M., Medvedev, O. S., *Pharmocological Regulation of Emotional Stress.* Moscow, 1979. pp. 360.

Valdman, A. V., Poshivalov, V. P., *Pharmocolological Regulation of Intraspecific Behavior.*

Valdman, A.V., Adaptive Restructuring of the Brain's Neural System with Extended Application of Antidepressants. *Neuronal Basis of Psychotropic Effects.* Edited by Academician of the USSR Academy of Medical Sciences A.V. Valdman. Moscow, 1982. pp. 8-32.

Valsilev, V.N., Chugunov, V.S., *Sympathetic -Adrenal Activity in Consideration of the Distinctive Functional Conditions of Man.* Moscow. 1985. pp. 272.

Valsilev, V.N., *Sympathetic-Adrenal Activity Considering the Distinctive Functional Conditions of man.* Abstract from Doctoral (MD) Dissertation. Moscow, 1981. pp. 36.

Vartapetov, B. A., Gladkova, A. I., New data on the Complex of Therapy for Post-Catastrophic Disorders. *Physiology, Biochemistry and Pathology of the Endocrine System*: Materials of the Rep. Conf. Kiev, 1971. pp 19-21.

Vasilev, V.N., Chugunov, V.S., Trial with 0.1 g. of L-Dopa in Clinical Neuroses and Neurosis-Like Conditions. Questions of Medical Chemistry 1983 Vol. 29, no. 3 pp. 119-124.

Bibliography

Vedernikova, N.N., Maykovsky A. I., Opiates and Endogenous Morphinic Peptides: A Systematic Approach to the Appraisal of their Roles in the Integration of the Nervous and Endocrinological Regulation in the Organism. *Successes of Modern Biology.* 1981. Vol. 91, abst 3. pp. 380-392.

Vedyaev, F. P., Volrobyova, T. M., *Models and Mechanisms of Emotional Stresses.* Kiev, 1983. pp. 136.

Veropotvelyan, P. N. Zubkov, G. B., Serinenko N. G., Kruchova, L. P. The Effects of Serotonin on the Neural Secretions of the Hypothalamus. *Journal of Experimental and Clinical Medicine.* 1980. Vol. 20, no. 6. pp. 600-603.

Veshchilova, T. P., Effect of Combined Application of Steriodal Preparations with Pyroxene on the Gonadotropic Function of the Pituitary. *Obstetrics and Gynecology.* 1975. no. 2 pp. 10-12.

Viru, A.A., *Central Nervous Regulation of the Stress Reaction of the Pituitary-Adrenalcortical System.* Academic Publication of the Western Tartar University. Tartu 1978. Vol. 8. pp. 3-53.

Vorontsov, V.I., *Characteristics of the Metabolism of Corticosteriods of lower monkeys in Normal and Stressful Situations.* Abstract from Doctoral Dissertation of Biological Sciences. Leningrad, 1972. pp. 22.

Zabrodin, O.H., Lebyedev, S. N., The Effect of Neurotropical Agents on the Healing of Erosion of the of the Muco-Membrane and the Contents of the Noradrenal in the Walls of Mouse Stomaches. *Pharmocology and Toxicology.* 1979. Vol. 42, no. 5. pp. 484-487.

Zabrodin, O.H., *The Role of Adrenergic Mechanisms in the Development and Healing of Experimental Neurological Injury Mucal Lining of the Stomach (A Pharmocological Analysis):* Abstract from the Dissertation for Doctorate of Medical Sciences. Leningrad 1982. pp. 142.

Zarechnii, V. R., *Cathecholiminals in the Wall of the Blood Carrying Veins of Mammals in the Process of Ontogological Development:* Abstract from the Dissertation of Doctorate of Medical Sciences. Leningrad 1979. pp. 24.

Zavodskaya, I. S., Moreva, e. V., Novikova, N. A., *The Effects of Neurotropical Agents on Neuro Heart Damage.* Moscow. 1977. pp. 192.

Zilye, R.K., Kusha, V.E., Correlation between the Neural-Methodology of the Brain and Hypothalmic Hormones. *Prob. Endocrinology.* 1979. Vol. 25. no. 6. pp. 73-79.

Zonis, B. J. Brin, V. B., Changes of Cardio-and Hemodynamics under the Cure of a-Adrenal-Pituitary Blockage of Patients with Arterial Hypertonia. *Terapevt. Archives* 1977. Vol. 49, no. 11. pp. 28-30.

SUBJECT INDEX

17-oxypregnenolone, 73

adaptagens, 155
adaptation, 46, 89, 92, 119
  period, 28, 106
  to confinement, 101, 120
adrenalin, 132
adrenergic
  mediation, 23, 37, 148
  reception, 33
  structures, 96
  structures of the brain, 45
adreno-cortical
  activity, 52, 55, 65, 73, 77
  reaction, 16, 74, 104, 113
  response, 62
AG in gonadectomized animals, 113
aggressive behavior, 37
AKTG, 59, 85, 124
  influence, 58
amygdala, 13
analogous stress stimulant, 31
androgen production
  in men, 150
  under stress, 86
androgenopoesis, 83
androgens, 18
anti-depressants, 45
anti-stress action, 153
arterial
  hypertension, 96, 117
  pressure, 106, 107, 149
athletes, 33, 37
aversive, 13

bioelectrical activity of the cortex, 13
biogenic amines, 17, 120, 121, 144
blockade of A-adreno receptors, 60
braking stimulant, 61

C-rhythm of adreno-cortical activity, 52
cages, 101
Calm, 47, 89, 124, 126, 130, 132
cannulation of the testical vein, 80
cardiovascular
  system, 105, 110
  illnesses, 150
Catecholamine excretion, 34, 97, 130
  in urine, 30, 92
catecholamine, 18, 85, 133
  reserves, 131
  in urine, 99
change in testosterone content in the blood, 87
changes in neuroticization, 13
chorionic gonadatropin, 89, 90, 91, 132
chronic
  psychogenic factors, 79
  psychogenic stress, 65
  stress, 37, 45, 64, 71, 96, 100
CKG, 89, 92
  under stress, 94
concentration of testosterone, 80

confinement, 92, 97, 101, 119
  in individual cages, 105
cortico-steroids, 56

DA
  discharge, 43
  excretion, 31
    in urine, 49
daily rhythms of gluco-corticoid secretion, 52
depressive-type state, 147
development of the stress state, 17
dofamine, 117
dogs, 33
dynamics of the neuro-hormonal indicators, 39

electro-shock therapy, 38
embryogenesis, 117
emotional tension, 44
emotional-stress stimulants, 12
endocrine function of the reproductive system, 80
endocrine function of the testes, 79, 86, 90, 113, 118, 150
endocrine reaction of the testes, 88
endocrine system, 11
evening hours, 30, 52, 83, 120, 122
excretion of KA, 104
excretion of NA, 43, 98
exogenic luliberin, 118
exogenous hormones, 59
experimental arteriosclerosis, 117
experimental neuroses in monkeys, 12
extreme emotional tension, 9
extreme stress, 50

fall, 52
frequent stress situations, 39
GGAKF, 62, 64
  activation, 59
GGAKS, 16, 74, 76, 83, 85, 88, 128
  reaction to stress, 53
  under stress, 115, 133
gluco-corticoid hormones in Monkeys, 51
gluco-corticoids, 18
gonadatropic function of the pituitary, 89
gonadectomy, 80
gonadotropic function of the pituitary, 81
group-hierarchic position, 13

heart rhythm, 106
high values of AD, 105
hippocamp, 13
hormonal
  activity of the testes, 83, 84, 85
  balance, 131
  reactions, 56
hormone concentration, 88
  in the blood, 52, 103
humoral-hormonal relationship, 115
hydrocortisone, 53, 61, 65, 66, 70, 71, 75, 101, 112, 122, 129
  concentration in the blood, 63, 69, 126
  content, 119
    in the blood, 104
hypokinesia, 64
hypothalamic regulation, 137
hypothalamus, 13
hypothalamus-hypopituitary-adrenal activity, 58
hypothalamus-pituitary-gonad system, 88

immobilization, 35, 44, 72, 73, 86, 98, 105, 122
  during the evening, 83
  leading to a significantly large increase in the level of hydrocortisone, 69
interhormonal
  interrelationships, 148
  relationship, 127
interrelation between emotions and KA secretion, 36

KA, 22, 111
  biosynthesis, 32
  exchange, 37
  excretion, 25, 45, 122, 124
  excretion, 50
  injection, 44
  synthesizing enzymes, 100
KA-ergic neurons, 117
KA-metabolizing enzymes, 117
KRF, 59

L-DOPA, 126, 128
  in Monkeys, 47
length of immobilization, 36
length of the stimulant, 33
lengthy stressor influences, 33
level of cortico-steroids, 59
LG concentration in the blood, 84
Luliberin, 89, 90, 91, 117, 132

male sex glands, 79
metabolic cages, 16
metopirone, 129
model emotional-stress states, 9
modelling of emotional stress, 7
monkey's brain, 10
morning hours, 30, 52, 53, 83, 121, 147

neurmediator systems, 116
neuro-emotional
  stimulant, 32
  strain, 57
  stress overload, 35
neuro-endocrine systems, 77
neuro-humorous regulating mechanisms, 13
neuro-peptide systems, 119
neuroendocrine
  components, 95
  impulses, 135
    impulses under ES, 19
  mechanisms, 19, 86
  system, 15, 89, 95, 117
neurogenic AG, 111
  arterial hypertension, 148
  illnesses, 7
neurohormonal balance, 152
neuropeptides, 136, 144
neurosis, 58
  in lower monkeys, 12
neuroticize monkeys, 12
neurotrophic preparations, 136
norepinephrine, 117, 132

Papio Hamadryas, 11
peripheral adrenergic mechanisms, 118
persistent arterial hypertension, 113
physical loads, 33
pituitary-adrenal complex, 128
pituitary-adreno-cortical system, 137
pituitary-testicular system, 93
preliminary immobilization, 77
preliminary stress, 39
prolonged emotional stress, 63
pronounced emotional stress, 44
protein complexes, 117
psycho-emotional sphere, 119
psychoendocrine research, 16
psychogenic stimuli, 37
Pyrroxan, 135

radioligand methods of
    analysis, 118
rats, 45
regulation of gluco-corticoid
    secretion, 58
repeated emotional stress, 65, 66,
    67, 73, 87, 97, 99, 101, 102,
    104, 105, 107, 109, 123, 148,
    152
repeated immobilization, 45
repeated influence of stressor-
    stimuli, 37
repeated stress irritants, 64, 76,
    112
resistance
    to emotional stress, 10
    to ES, 89
response reaction of SAS to stress
    stimulants, 24

SAS activity, 28
SAS's extreme sensitivity, 23
schizophrenics, 38
secretion of NE and A, 37
selectivity of the disturbance, 15
self-regulation of the
    physiological functions, 23
sensitivity of the testes to
    endogenic LG, 84
serotonin, 118
serotonin-ergic system, 116
severe emotional stress, 103
severe emotional-pain stress, 24
severe stress, 43, 83, 91
sex glands, 81, 89
sex hormones, 118
single-phase C-rhythms, 81
somato-visceral
    function, 13
    systems, 14
spring months, 52
steroid hormone content in the
    blood, 119

steroid
    hormones, 17, 120, 121
    in the blood, 126
    producing glands, 18, 119, 146
    stimulation by a tropic hormone,
        93
stress
    application-immobilization,
        70
    during the evening, 31
    injuries to internal organs, 46
    irritants, 42
    stimulants, 25
    stimulation application, 74
stress-realizing systems, 46
structural-functional
    organization, 14
substitute activity, 13
suppression of
    androgenopoesis in the
        testes, 148
    the endocrine function of the
        testes, 86
swimming, 52
sympatho-adrenal system, 117,
    120, 146, 147
systolic pressure, 105

testes under stress, 80
testosterone, 79, 83
    concentration in the blood,
        103
    content in blood, 82, 83, 90,
        94, 101, 119, 153
    in the blood under stress, 85
    level in the blood, 91
time of day, 18

vanilil-phenyl-glycolic acid, 132
vascular tension, 110
vegetative index, 108
visceral systems, 15
VMK excretion, 43, 44
VND normalization, 28